ESHWATER F SH AND IRELAND, OTES ON THEIR DI TION AND ECOLOGY

by

PETER S. MAITLAND*

F SHWATER BIOLOGICAL ASSOCIATION
SCIENTIFIC PUBLICATION No. 62

2004

Editor: D. W. SUTCLIFFE

*Fish Conservation Centre, Gladsh EH41 4NP, Scotland

The Environment Agency welcomes the publication of *Keys to the freshwater fish of Britain and Ireland, with notes on their distribution and ecology*.

When the water industry was embarking upon serious fisheries science in the early 1970s, the FBA's freshwater fish key by Peter Maitland (No. 27, 1972) was the definitive work when learning the basics of fish identification and deciding just exactly what fish was in the bucket! Inevitably, copies became difficult to acquire and those who had not worn out their originals by frequent use became reluctant to let them out of sight. Many new to fisheries work in the Environment Agency thus have no knowledge of the original publication. Thirty or so years on, it is a delight to see the new version available, fully revised, to once again become the definitive document for identifying the freshwater fish of Britain and Ireland. The new book will sit nicely alongside the recent FBA *Keys to larval and juvenile stages of coarse fishes from fresh waters in the British Isles*, by Adrian Pinder.

The production of this new volume could not be more timely. These days, graduate recruits to the Agency tend to be generalists rather than specialists whilst at the same time the need for fish monitoring is being driven to an increased level of importance by legislation such as the Water Framework Directive. Also, the Environment Agency places much emphasis on 'a better quality of life' and 'an enhanced environment for wildlife', for which the understanding and management of fish populations are fundamental components. Peter Maitland's new keys, particularly in relation to the less common and more difficult species, will be an essential tool for fisheries and environmental scientists, both within the Agency and outside it. The series of FBA keys to freshwater fauna are renowned for their quality and usefulness, and the Environment Agency is pleased that such a quality aid to freshwater fish identification is once again published in the FBA series of Scientific Publications.

Phil Hickley
National Fisheries Technical Manager
Environment Agency

Published by the Freshwater Biological Association,
The Ferry House, Far Sawrey, Ambleside, Cumbria LA22 0LP, UK

Freshwater Biological Association 2004

ISBN 0–900386–71–1

ISSN 0367–1887

PREFACE

Since Peter Maitland's previous key was published as FBA Scientific Publication No. 27 in 1972, the freshwater fish fauna of Britain and Ireland has changed quite considerably, both in composition and distribution. Three new species have been introduced through the trade in ornamental fish, and two of the species formerly recorded are, for all intents and purposes, now extinct in the British Isles. The distribution of existing species has also changed considerably, through introductions to new catchments or reductions through man-induced habitat degradation. At the same time fishery management has intensified with an increased popularity of angling and a greater awareness by regulatory authorities, owners, conservationists and amateur naturalists, of management requirements in order to achieve sustainable fish communities.

Over the last 30 years many books have been published to assist in the identification of fish species, both marine and freshwater. They range from large, international electronic compendia of up to 25,000 species, together with their descriptions, biology and ecology, to the most popular illustrated books, with many combinations in between.

However, the extensively revised text in this new publication provides a comprehensive set of keys to adults, fry, eggs and scales, and is illustrated with both line-drawings and photographs. It contains the most up-to-date notes on distribution and ecological information in an easily assimilable format and some 700 references are listed for those wishing to pursue more detailed aspects.

Keys to the freshwater fish of Britain and Ireland complements the recently published *Keys to larval and juvenile stages of coarse fishes from fresh waters in the British Isles* by Adrian Pinder (FBA Scientific Publication No. 60, 2001). Publication of this sixty-second volume, the latest in our series of Scientific Publications, has been made possible by the enthusiasm and total commitment of the author, Professor Peter Maitland, and collaboration from the Environment Agency which has provided generous sponsorship. Northumbrian Water and Thames Water have also helped with sponsorship. We are most grateful for the interest of these organisations, which are keen to demonstrate their ecological awareness.

The Ferry House
December 2003

Roger Sweeting
Chief Executive, FBA

CONTENTS

INTRODUCTION

The keys presented here are a revised version of those given by Maitland (1972a), and provide a simple means of accurately identifying any fish which occurs regularly in fresh water anywhere in Britain and Ireland. There is a large literature covering various aspects of the biology of the freshwater fish of the British Isles that will not be reviewed here. However, many of the more important publications, some of which contain useful bibliographies, are included in the list of references on pages 209–239. Some of the most relevant references for each of the 57 species dealt with in this publication are also listed. Classical accounts of our ichthyofauna include those of Yarrell (1859), Houghton (1879), Maxwell (1904), Regan (1911) and Jenkins (1925). Recent guides to European freshwater species include those of Wheeler (1969), Miller & Loates (1997) and Maitland (2000). Many of the fish concerned can be identified directly by comparison with illustrations in some of these publications, but this is often a slow and not always an accurate method. Moreover, most of the older works do not include species that have been introduced recently and become established, such as the Sunbleak *Leucaspius delineatus* and the False Harlequin *Pseudorasbora parva*.

The taxonomic classification used in this publication largely follows that of Nelson (1994), Kottelat (1997) and Maitland (2000).

All fish in the British Isles, apart from lampreys, are true bony fish or teleosts. Lampreys, although they have no proper jaws and thus are technically not true fish, are included in the keys, together with those fish which, while mainly marine, are also found regularly in fresh and brackish waters. Marine species that are rarely found in brackish waters have been excluded from the keys, although a few are mentioned that might be confused with fresh- or brackish-water species. Most of our freshwater fish are native to these islands, but several introduced species that have become established here are included. Others, whose status is often doubtful and usually temporary, have not; among them are various tropical species associated with heated effluents (e.g. Meadows 1968) and temperate species for which there are odd records but no evidence of the existence of an established population (see below).

In any assessment of the fish fauna of the British Isles it is impossible to ignore the effects of history and geomorphology. Probably the most important event over the last 30,000 years was the development of the ice sheet which completely buried most of the mainland and surrounding island groups. This last glaciation came to a close some 10,000 years ago and can be taken as the starting point for any consideration of the present freshwater fish fauna of Britain and Ireland (Maitland & Campbell 1992).

After the last glaciation there was a relatively easy colonisation of the fresh waters of these islands from the sea by those fish species with marine affinities, followed by a slow natural dispersal and an increasingly faster rate of transfers, by humans, of other native species – purely freshwater – from the south. Additionally, alien species from abroad have been introduced. All of this has resulted in a somewhat north/south and to a lesser extent east/west divide in the distribution of species and the composition of fish communities in the British Isles. The situation is more complex than dividing lines on a map, but it is certainly true to say that the fish communities in the north and on islands have many fewer species than comparable systems in the south.

Thus, for example, typical northern running waters in the north of Scotland like the Rivers Naver, Thurso and Helmsdale have only about six or seven species of freshwater fish, whereas comparable waters in southern Scotland, for example the Rivers Nith, Annan and Tweed, have sixteen to nineteen species. Further south still, in England, the Rivers Severn, Avon (Hampshire) and Thames may have twenty-five to thirty species. As with running waters, fish communities in lakes in the north and on islands have many fewer species than comparable systems in the south. Thus, for example, typical northern standing waters, such as Lochs Stack, Shin and Calder have only six or seven native species whilst Lochs Ken and Lomond and St Mary's Loch in the south of Scotland have nine (Ken) to fifteen (Lomond). In the south of England, several additional native species (e.g. Silver Bream *Abramis bjoerkna*) increase the complexity of communities. In terms of species composition, many of the northern communities probably have remained the same for thousands of years and are relatively stable, whereas those in the south have changed substantially over the last two centuries due to human impacts and continue to do so with more recent introductions by anglers and others.

Distribution maps (pp. 180–207) show the general occurrence, by Hydrometric Areas (see pp. 177–179), of each species of fish found in Britain and Ireland. More detailed maps of distribution in Britain (not Ireland) are available in Davies et al. (2004). The distribution of many species is changing; new information would be welcome and should be passed to appropriate local museums or biological record centres.

Colour illustrations (pp. 89–112) cover most of the species found in the British Isles. They are not displayed in scientific order but in groups of species that may look similar, in order to aid final identification.

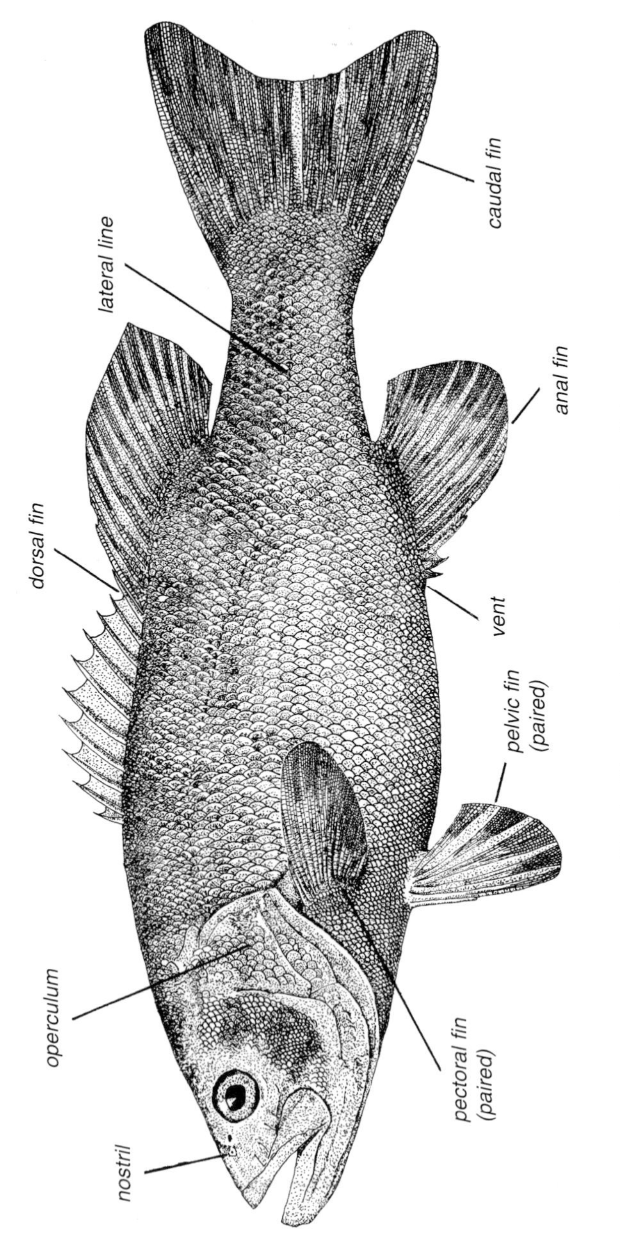

Fig. 1. General external features of a bony fish.

FISH STRUCTURE

The basic external features of a bony fish are shown in Fig. 1.

Sense organs. In all fish, most of the obvious external sense organs are located on the head: a pair of eyes, the nostrils (normally paired, except in lampreys), and often barbels that may vary in number, size and position according to species. Running along each side of the body in most fish is the lateral line – a long sensory canal just under the skin but connected to the exterior by a series of tubules. These often run through individual scales. The main function of the lateral line is sensory and its nerve cells are able to pick up very fine vibrations in the water. Branches of the same system run on to the head region. The extent and position of the lateral line varies with species. In most fish it runs laterally from head to tail; in some species it runs midway along the body, in other species it is curved in a particular way. In a few species the lateral line terminates only a short distance from the head, or is absent.

Mouth region. That part of the head anterior to the mouth is normally termed the snout. The position of the mouth itself may be terminal, inferior (opening downwards) or superior (opening upwards); in a few species of fish it is modified to form a sucker.

Bones and teeth. Associated with the mouth are several bones (Fig. 2, p. 10) that are of taxonomic importance (e.g. maxillary, premaxillary, vomer, hyoid, palatine); several of these may carry teeth which can be long or short, permanent or deciduous. In adult lampreys, oral discs are present, which have anterior and posterior areas bearing teeth. The mouth opens into the pharynx at the back of which there are, in some fish, bony structures specialised for chewing and crushing. These are the pharyngeal bones (Fig. 3, p. 15) and their structure is important in confirming the identification of many cyprinid fishes. Pharyngeal bones are modified fifth gill arches, whose structure is described later (p. 25).

Head region. The brain is well protected inside the bony skull, at the back of which, also enclosed in bone, are the semicircular canals (inner ears) which help the fish to maintain its balance. Within a chamber, inside each canal, loose pieces of calcium carbonate are secreted. These otoliths, like scales, grow in proportion to the size of the fish and have growth 'rings' which can be used when ageing specimens. They also vary with species and can be used in identification – less important when fresh fish with many other characters are available, but useful in scatological and archaeological studies when they may be the only part of the fish that has not been digested or decayed.

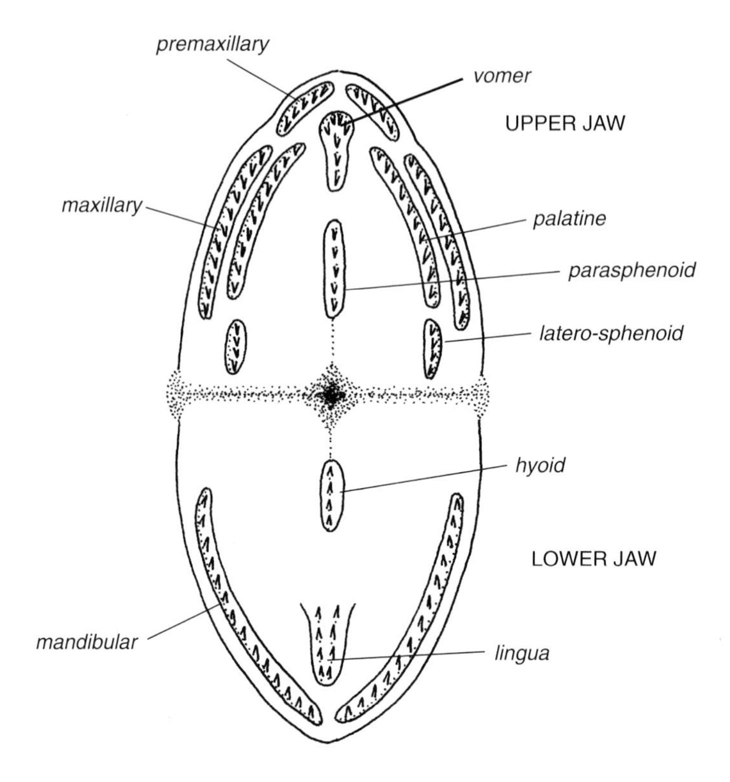

Fig. 2. Diagram of the main bones associated with the mouth of a bony fish.

Gut. The pharynx leads into the oesophagus which opens, in most fish, to the stomach. Note, however, that the stomach is absent in some families (e.g. cyprinid fish). Food is held in the stomach for some time before being passed into the intestine where it is digested; undigested materials pass on into the rectum, from which they pass out through the anus as faeces. The anus and the genital opening share a common space, known as the cloaca, which opens to the outside through the vent. Most fish are able to control their buoyancy in the water by means of a swim bladder which is derived from the gut and is usually situated above it. In some groups of fish (e.g. Cyprinidae) the swim bladder is connected to the oesophagus by a pneumatic duct.

Gills. At the sides of the pharynx are cavities leading past the main respiratory organs, the gills; together these form the branchial region. Each gill consists of a strong supporting arch, on one side of which is a set of comb-like rakers whose function is to prevent material from the pharynx entering the delicate blood-filled respiratory lamellae which are aligned on the other side of the gill arch. In most fish there are four gills on either side, the passages between them leading to the outside of the body through the gill openings. The latter are protected on each side by a single bony gill cover known as the operculum, except in lampreys where the gill openings lead directly to the outside.

Scales. The whole body is covered by skin and within this in most fish lie small bony plates known as scales, forming a protective and flexible covering over most areas except the head, which is protected by the head bones. The number and structure of the scales varies from species to species and thus they are useful for identification purposes (see p. 121). Some species have no scales; in others the scales are replaced by isolated bony scutes which project from the skin. Scales along the lateral line are characterised by having narrow tubules running through them. Within the skin also there are pigment cells of various kinds, responsible for much of the colouration of the fish. Though colour patterning can be useful for distinguishing various species of fish, it should be regarded with caution, for even within a single species the colour can vary greatly with age, sex, season, time of day, or even emotional state.

Fins. The typical arrangement of fins on a fish is shown in Fig. 1. There are two sets of paired fins, both situated ventrally: the pectoral fins and the pelvic (or ventral) fins. These are respectively equivalent to the fore-legs and hind-legs of terrestrial vertebrates. On the back is a dorsal fin; in some species this may consist of two distinct parts, or be divided into two separate fins, or have the anterior of these represented by several isolated spines. Behind the dorsal fin in the Salmonidae and related families there is a small fleshy fin with no rays, the adipose fin. The portion of the body posterior to the vent is known as the caudal region. Ventrally, just behind the vent this bears the anal fin whilst further back, behind where the body narrows to form the caudal peduncle, is the caudal (or tail) fin. The supporting structures of fins are known as rays; these may be branched or unbranched (when they are usually referred to as spiny or bony) and are often useful taxonomic characters.

COLLECTION AND PRESERVATION

Fish may be caught by a wide variety of methods depending on the species concerned, its size, habitat, etc. and, of course, the legality of the method. The latter may vary in different parts of the country and even seasonally, where close periods operate. Some fish are given special protection nationally through the 1981 Wildlife and Countryside Act (e.g. Vendace), which is discussed on page 18. Among the most common methods of capture are angling for larger fish and the use of a handnet for smaller fish. Even a handnet may require a licence in some places and the user must always check that permission is available and that a licence, where necessary, is held. There are more efficient methods of capture (some of them much more efficient but illegal unless under licence) which can be used, e.g. electro-fishing, gill nets, seine nets, trawl nets and traps. These are described in detail elsewhere (Maitland 1990) and will not be discussed here. Most methods of fishing are highly selective and may capture only one size-group of one species, and sometimes even only one sex. In carrying out any detailed study of several species in a water with mixed habitats it is normally advisable to use several different methods of capture.

The present key will not serve to identify the very small specimens (larvae and fry) of most species, but for coarse fish a suitable key has become available (Pinder 2001). For a short time after hatching, the young of most species change rapidly in form, and features characteristic of the adults do not appear for some time. Unlike many invertebrates, adult fish can often be identified in the field immediately after capture and they can then be returned to the water alive. There is little point in killing fish unless they are required for food, research or some other legitimate purpose. Species that are difficult to identify in the field may have to be taken away for careful examination and accurate identification. For this purpose, it is preferable that they are kept alive (and subsequently returned to the wild), but this may not always be possible, especially with large or delicate specimens, and such fish may have to be killed by some means or other. Clearly all specimens, where dissection is essential for their accurate identification, unfortunately must be killed.

One of the most humane ways to kill fish is to use a liquid anaesthetic such as MS222 (note: like formaldehyde, this chemical has carcinogenic properties and must be handled with care – see below). Specimens dropped into a 0.1% solution of this are narcotised very quickly and can then be frozen or transferred to a suitable fixative. Fish should be examined as soon as possible, ideally within 48 hours, or sooner in warm weather. If they cannot be examined within this time they must be frozen or fixed in some way.

Frozen fish keep their true colours better than fixed fish. The best procedure is to place each specimen in a polythene bag with a little water and freeze the bag and contents as quickly as possible. The fish should be kept straight during this process and care must be taken not to damage fins – especially the tail fin, which is vulnerable if material is being moved about within a deep freeze refrigerator.

Even frozen fish will not keep indefinitely, however, if they have to be subjected to periodic thawing for examination purposes, and normally specimens to be stored must be fixed permanently in some way. The traditional fixative is 4% formaldehyde, though great care must be taken when using this chemical which is now known to have carcinogenic properties. Ideally, each fish should be preserved by placing it flat on its side in a shallow dish with its fins spread as much as possible and then pouring enough of the fixative over it to cover it completely. Specimens should be left for several days to ensure complete fixation and, where formaldehyde is being used, the whole procedure should be carried out in a well ventilated place – ideally a fume cupboard in a laboratory, or outside or in an outbuilding if done at home. It should be noted that, after preservation, most fish shrink slightly, resulting in the length of a preserved fish being less than when it was fresh.

With large fish (more than 30 cm in length) it is advisable to make a small slit in the ventral body wall, or to inject the body cavity with a small amount of 40% formaldehyde to ensure complete fixation internally. The fish can then be stored temporarily in strong polythene bags or more permanently in suitable jars, either in 4% formaldehyde or in more pleasant preservatives such as 70% alcohol or 1% propylene phenoxetol (Owen 1955). Each container should have inside it a suitable label, written in soft pencil or indelible ink, with a note of the species concerned, the water where it was collected (with a National Grid Reference to confirm the site), the method of capture, the date of collection and the name of the collector.

In situations where the suggested preservatives are not available, quite good results can be obtained with more easily obtainable materials such as: (a) methylated spirits, diluted seven parts of spirits to three parts of water, (b) a saline solution of one part of salt mixed with two parts of water, (c) vinegar (acetic acid) as used as a condiment. Preservation can be carried out satisfactorily in polythene bags, keeping the fish as straight as possible. Ideally, valuable specimens should be transferred to 4% formaldehyde as soon as possible.

Eggs or larvae can be preserved and stored in small tubes containing either 4% formaldehyde or 70% alcohol. Labels with relevant data (species, locality, date, colour of eggs when fresh, exact habitat, name of collector) should be placed inside each tube. With eggs, a reasonable number should be taken if possible, especially if they are adhering to each other, for the form of attachment may be important for identification.

Some species of fish can be identified from their scales alone, and in addition it is often possible, with several good scales, taken from the side of the body (see Fig. 1, p. 8), to establish the specimen's age and certain features of its past history. Scales should be placed inside a small envelope which is then flattened and allowed to dry; they will keep indefinitely in this way. On the outside of the envelope the following data should be recorded: species, locality, date, name of collector, and length, weight and sex of the specimen concerned. The length of the fish can be determined in a number of ways (Maitland 2000): the most useful method is to record the exact distance between the tip of the snout and the tip of the middle ray of the tail fin – known as fork length. For examination, the scales should be carefully cleaned by wetting again and gently rubbing individual scales between the thumb and forefinger. They should then be mounted on a glass slide, either dry or in glycerine jelly.

As with other vertebrates, bones are often very useful in the identification of fish and in determining their age. In some cases (e.g. the pharyngeal bones of Cyprinidae and the vomer bones of Salmonidae) they may be almost essential for accurate identification. Certain bones are of major importance in providing information on the age and growth of some species (e.g. the opercular bones of Percidae and Esocidae). Preparation of such bones for examination and subsequent preservation and storage is a relatively simple task; fresh or frozen material should be used, not material that has been fixed. The relevant bones should be dissected out from the fish concerned along with any adhering tissues. Each bone should be dropped into very hot water for a few minutes and then scrubbed gently with a small stiff brush (e.g. an old toothbrush) to clean away soft tissue. The process should be repeated until the bone is completely clean; it can then be placed on clean paper and allowed to dry out slowly in a warm (but not hot) atmosphere. Details concerning the fish from which the bone was removed (species, locality, date, name of collector, and the length, weight and sex of the specimen) should be written on a small stiff label attached to the bone by a strong thread, or on the outside of an envelope or small box in which the bone is kept. Bones cleaned and dried in this manner will keep indefinitely.

Pharyngeal bones should be treated like other bones. To locate them in cyprinid fish, open the pharyngeal cavity ventrally, or from the side, and feel carefully for the teeth of the pharyngeal bones just behind the gill slits and in front of the cleithrum (pectoral girdle) (Fig. 3) (Sibbing et al. (1986); Lammens & Hoogenboezem (1991); Sibbing (1991)).

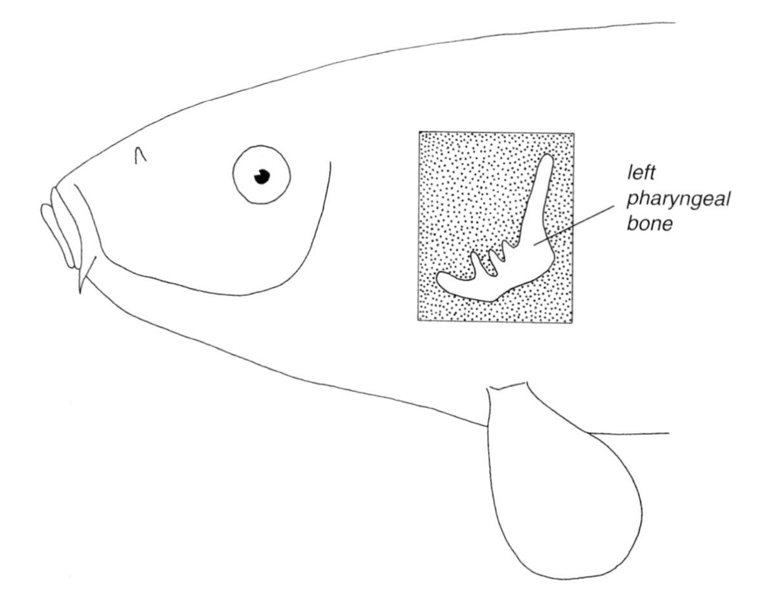

Fig. 3. Position of the left pharyngeal bone in a typical cyprinid (Tench *Tinca tinca*).

When the pharyngeal bones have been located, remove them carefully by cutting away any attachment tissue. In small fish, particularly, the teeth are very delicate and care must be taken not to break them. The 'ear bones' (otoliths) are easier to remove (for they have no attachment) but are more difficult to locate. To find them, cut vertically downwards into the skull, just behind each eye, when they will be found on each side lying within the sacs of the semicircular canals. They should be rinsed and then dried like bones.

A great many of the species of freshwater fish which occur in the British Isles can be kept quite successfully in captivity, either indoors in aquaria or outdoors in ponds. The provision of adequate living conditions is simple, the main requirements being reasonable space (with as large a surface area as possible), cool clean water, sufficient cover (in the form of aquatic plants or stones) and appropriate food for the species concerned. Species such as Common Minnow and Bitterling, which will happily feed on dried food, are generally easier to keep than, say, Three-spined Stickleback or Common Bullhead, which require live food. Further details are not given here for there are several admirable books that deal specifically with the subject of keeping fish.

LEGISLATION

Before the 1970s, the protection given to most native fish in Europe was inadequate in terms of management, legislation and the establishment of appropriate reserves (Maitland 1979; Maitland & Lyle 1991; Lyle & Maitland 1992). The exceptions mostly related to fish of angling importance which are given substantial protection by local management and through available legislation. However, since the early 1970s, there have been a number of international and national conventions and laws in Europe which have led to much better protection for freshwater fish and their habitats. The following are the main pieces of international legislation and anyone working with freshwater fish in Britain and Ireland should be aware of the protection given to certain species.

Ramsar Convention 1971. This was the first international conservation agreement for the protection of internationally important wetlands. By 2000, there were 1005 wetland sites designated by 116 Contracting Parties. These sites cover 71 million hectares. The main objective is the protection of habitat and although individual species are not protected as such, many of the sites have open waters and give habitat protection to their freshwater fish.

CITES 1975. The Convention on International Trade in Endangered Species (CITES) seeks to encourage governments to regulate, and in some cases prohibit, trade in species threatened with extinction. Several fish species are listed in the CITES appendices, but only one of these is relevant to Britain and Ireland – Common Sturgeon.

Bern Convention 1979. The Bern Convention on the Conservation of European Wildlife and Natural Habitats requires the protection of endangered and vulnerable species of flora and fauna in Europe and their habitats. Appendices list species for which exploitation and other factors should be controlled. No species in Britain or Ireland are listed in Appendix II (strictly protected species), but the following species all appear in Appendix III (protected species): River Lamprey, Brook Lamprey, Sea Lamprey, Common Sturgeon, Allis Shad, Twaite Shad, Bitterling, Spined Loach, Wels Catfish, Vendace, Pollan, Powan, Houting, Grayling, Atlantic Salmon and Common Goby.

Bonn Convention 1979 and 1994. The Bonn Convention on the Conservation of Migratory Species of Wild Animals requires the protection of migratory animals, from the Arctic to Africa. So far, no species of relevance to Britain and Ireland have been listed.

Convention on Biological Diversity 1992. The Convention on Biological Diversity was signed by 150 heads of state and government at the Earth Summit in Rio de Janeiro in June 1992. Following this, in the United Kingdom, the UK Biodiversity Action Plan was published in January 1994. The Action Plan set out a strategy for the next 20 years (Department of the Environment 1995). A vital part of the strategy are schemes for the conservation of certain endangered wildlife habitats and species of animals and plants. The aim of these schemes will be to preserve, and wherever possible enhance, the range and biodiversity of naturally occurring wildlife in the UK. The Priority List of species in the UK for which Species Action Plans have been prepared includes the following freshwater fish species: Allis Shad, Twaite Shad, Vendace, Pollan, Houting and Burbot.

Habitats Directive 1992. The European Union adopted the Habitats Directive (Directive 92/43/EEC on the conservation of natural habitats and of wild flora and fauna) in May 1992. The main aim of the Directive is 'to promote the maintenance of biodiversity, taking account of economic, social, cultural and regional requirements' (Department of the Environment 1995). The Habitats Directive is thus European Law which provides for the creation of a network of protected areas across the European Union to be known as 'Natura 2000'. These protected areas will consist of Special Areas of Conservation (SACs) designated under the Habitats Directive and the much older Special Protection Areas (SPAs) designated under the Birds Directive (Directive 79/409/EEC on the conservation of wild birds). The Habitats Directive also requires all Member States to set up an effective system to prevent the capture, killing, injuring or damaging disturbance of certain endangered species. Together, these measures aim to maintain or restore the extent and quality of rare habitat types and to ensure that rare species can survive and maintain their populations and range on a long-term basis. Freshwater fish are included in three Annexes in the Directive: II (conservation requiring Special Areas of Conservation), IV (species in need of strict protection), and V (taking in the wild and exploitation may be subject to management measures). Species relevant to Britain and Ireland which are included in this Directive are: River Lamprey (II, V), Brook Lamprey (II), Sea Lamprey (II), Common Sturgeon (II, IV), Allis Shad (II, V), Twaite Shad (II, V), Barbel (V), Bitterling (II), Spined Loach (II), Vendace (V), Pollan (V), Powan (V), Houting (II, IV), Grayling (Va), Atlantic Salmon (II, V), and Common Bullhead (II).

Water Framework Directive 2000. The Framework Directive for Water (2000/60/EC), commonly known as the Water Framework Directive (WFD), was published in December 2000. It identifies its key purpose as preventing further deterioration of, and protecting and enhancing the status of, aquatic systems in Europe. Member States will be required to achieve 'good surface water status' in inland surface waters, transitional waters and coastal waters. Ground waters must also be protected and restored to ensure the quality of dependent surface water and terrestrial ecosystems. The presence of appropriate fish communities in freshwater habitats is one of the main criteria in determining water quality status. Unfortunately, the WFD focuses on physicochemical impacts, with those related to fisheries or alien species being excluded from consideration.

Within Britain and Ireland there is also relevant national legislation, among which may be mentioned the following:

The Wildlife and Countryside Act 1981 and 1985. This is the main piece of conservation legislation for the United Kingdom, and for implementing EU conventions. It is intended to protect both species and sites of UK importance, and it is an offence to 'kill, injure or take any wild animal included in Schedule 5', to 'damage, destroy or obstruct any shelter or disturb any Schedule 5 animal there', and to 'possess any live or dead wild animal (or part of) included in Schedule 5'. The following freshwater fish species are listed in Schedule 5: Common Sturgeon, Allis Shad, Vendace, Powan and Burbot.

Prohibition of Keeping or Release of Live Fish (Specified Species) Order 1998. Many alien species of concern, some already established in the British Isles, are listed in this Order (which is at present being revised), applicable only in England and Wales, whereby 'it is an offence to keep or release any non-native fish or shellfish' specified in the Order without an appropriate licence. A similar Order is likely to be made for Scotland in the near future.

Salmon and Freshwater Fisheries Acts. There are numerous Acts relevant to England and Wales, Scotland and Ireland which are concerned primarily with the protection of Atlantic Salmon and Sea Trout – or more correctly the fisheries for these species. These are not covered in detail here.

CONSERVATION STATUS

One of the first tasks in preparing conservation plans for the fish fauna of any geographic area is a proper assessment of the conservation status of each species. This is essential in being able to give priority to those species under greatest threat and to the preparation of conservation management plans. Most countries now accept the IUCN (1994) definitions of threat, which are used in the accounts of distribution and ecology (pp. 133–177) and are summarised as follows.

Extinct (EX): 'When there is no reasonable doubt that the last individual has died.'
Extinct in the Wild (EW): 'When it is known only to survive in cultivation, in captivity or as a naturalised population (or populations) well outside the past range.'
Critically Endangered (CE): 'When it is facing an extremely high risk of extinction in the wild in the immediate future.'
Endangered (EN): 'When it is not Critically Endangered but is facing an extremely high risk of extinction in the wild in the near future.'
Vulnerable (VU): 'When it is not Critically Endangered or Endangered but is facing a high risk of extinction in the wild in the medium-term future.'
Lower Risk (LR): 'When it has been evaluated, does not satisfy the criteria for any of the categories Critically Endangered, Endangered or Vulnerable.'
Data Deficient (DD): 'When there is inadequate information to make a direct, or indirect, assessment of its risk of extinction based on its distribution and/or populations status.'
Not Evaluated (NE): 'When it has not yet been assessed against the criteria.'

CHECKLIST OF SPECIES IN BRITAIN AND IRELAND

The species listed below are (or were – a few are believed to be extinct here now) all native to Britain and Ireland, or have been introduced from abroad and are now established in some waters.

Lamprey family: Petromyzontidae	
Lampetra fluviatilis (Linnaeus 1758)	River Lamprey (Lampern)
Lampetra planeri (Bloch 1784)	Brook Lamprey
Petromyzon marinus Linnaeus 1758	Sea Lamprey
Sturgeon family: Acipenseridae	
Acipenser sturio Linnaeus 1758	Common Sturgeon (Atlantic Sturgeon)
Eel family: Anguillidae	
Anguilla anguilla (Linnaeus 1758)	European Eel
Herring family: Clupeidae	
Alosa alosa (Linnaeus 1758)	Allis Shad
Alosa fallax (Lacepede 1803)	Twaite Shad
Carp family: Cyprinidae	
Abramis bjoerkna (Linnaeus 1758)	Silver Bream
Abramis brama (Linnaeus 1758)	Common Bream
Alburnus alburnus (Linnaeus 1758)	Bleak
Barbus barbus (Linnaeus 1758)	Barbel
Carassius auratus (Linnaeus 1758)	Goldfish
Carassius carassius (Linnaeus 1758)	Crucian Carp
Cyprinus carpio Linnaeus 1758	Common Carp
Gobio gobio (Linnaeus 1758)	Common Gudgeon
Leucaspius delineatus (Heckel 1843)	Sunbleak (Belica)
Leuciscus cephalus (Linnaeus 1758)	Chub
Leuciscus idus (Linnaeus 1758)	Orfe (Ide)
Leuciscus leuciscus (Linnaeus 1758)	Dace
Phoxinus phoxinus (Linnaeus 1758)	Common Minnow
Pseudorasbora parva (Temminck & Schlegel 1842)	False Harlequin (Topmouth Gudgeon)
Rhodeus sericeus (Bloch 1782)	Bitterling
Rutilus rutilus (Linnaeus 1758)	Roach
Scardinius erythrophthalmus (Linnaeus 1758)	Rudd
Tinca tinca (Linnaeus 1758)	Tench

Spined Loach family: Cobitidae

Cobitis taenia Linnaeus 1758	Spined Loach

Stone Loach family: Balitoridae

Barbatula barbatula (Linnaeus 1758)	Stone Loach

American Catfish family: Ictaluridae

Ameiurus melas (Rafinesque 1820)	Black Bullhead

European Catfish family: Siluridae

Silurus glanis Linnaeus 1758	Wels Catfish

Pike family: Esocidae

Esox lucius Linnaeus 1758	Pike

Smelt family: Osmeridae

Osmerus eperlanus (Linnaeus 1758)	Smelt (Sparling)

Whitefish family: Coregonidae

Coregonus albula (Linnaeus 1758)	Vendace (Cisco)
Coregonus autumnalis (Pallas 1776)	Pollan (Arctic Cisco)
Coregonus lavaretus (Linnaeus 1758)	Powan (Schelly, Gwyniad)
Coregonus oxyrinchus (Linnaeus 1758)	Houting

Salmon family: Salmonidae

Oncorhynchus mykiss (Walbaum 1792)	Rainbow Trout
Salmo salar Linnaeus 1758	Atlantic Salmon
Salmo trutta Linnaeus 1758	Brown Trout (Sea Trout)
Salvelinus alpinus (Linnaeus 1758)	Arctic Charr
Salvelinus fontinalis (Mitchill 1814)	Brook Charr

Grayling family: Thymallidae

Thymallus thymallus (Linnaeus 1758)	Grayling

Cod family: Gadidae

Lota lota (Linnaeus 1758)	Burbot

Grey Mullet family: Mugilidae

Chelon labrosus Risso 1826	Thick-lipped Grey Mullet
Liza aurata (Risso 1810)	Golden Grey Mullet
Liza ramada (Risso 1826)	Thin-lipped Grey Mullet

Stickleback family: Gasterosteidae	
Gasterosteus aculeatus Linnaeus 1758	Three-spined Stickleback
Pungitius pungitius (Linnaeus 1758)	Nine-spined Stickleback (Ten-spined Stickleback)
Sculpin family: Cottidae	
Cottus gobio Linnaeus 1758	Common Bullhead
Bass family: Moronidae	
Dicentrarchus labrax (Linnaeus 1758)	Sea Bass
Sunfish family: Centrarchidae	
Ambloplites rupestris (Rafinesque 1817)	Rock Bass
Lepomis gibbosus (Linnaeus 1758)	Pumpkinseed
Micropterus salmoides (Lacepede 1802)	Largemouth Bass
Perch family: Percidae	
Gymnocephalus cernuus (Linnaeus 1758)	Ruffe (Pope)
Perca fluviatilis Linnaeus 1758	European Perch
Sander lucioperca (Linnaeus 1758)	Pikeperch (Zander)
Goby family: Gobiidae	
Pomatoschistus microps (Kroyer 1838)	Common Goby
Flatfish family: Pleuronectidae	
Platichthys flesus (Linnaeus 1758)	Flounder

Species that are not included in the checklist

Only fish that are known to have been regular visitors, or have (or have had) established populations in the British Isles, or are introduced regularly, are included in the above checklist. Several non-native species that have been recorded recently in Britain or Ireland, but are assumed to be only casual individuals or fish that have been artificially maintained, are not covered in detail. These include the following ten species.

Siberian Sturgeon *Acipenser baeri* Brandt 1869:– stray specimens, probably from Russian releases in the Baltic or fish farm escapes, have been taken in the sea around the British Isles.

Sterlet *Acipenser ruthenus* Linnaeus 1758:– a single dead specimen was taken from the River Thames some years ago, probably dumped there by an aquarist. Specimens have been found recently in other rivers.

Grass Carp *Ctenopharyngodon idella* (Valenciennes 1844):– a number of waters in different parts of Britain have been stocked with this species in an attempt to control vegetation. It appears to be impossible for this species to breed here and so there are no established populations. However, it is probably the most widespread of those species that are stocked but have never become established, and a considerable amount of research has been carried out on its biology (Cross 1969, 1970; Edwards 1973; Stott & Cross 1973; Van Dyke & Sutton 1977; Colle et al. 1978; Kilambi & Robison 1979; Bettoli et al. 1985).

Guppy *Poecilia reticulata* Peters 1859:– formerly established in the vicinity of a heated effluent running into the River Lee in Essex. Now extinct with the demise of the business concerned.

White Sucker *Catostomus commersoni* (Lacepede 1803):– a single specimen taken on the River Gade at Hemel Hempstead, Hertfordshire. This was probably an escape from a fish farm just upstream where White Suckers were accidentally imported with other species (Copp et al. 1993).

Pink Salmon *Oncorhynchus gorbuscha* (Walbaum 1792):– several specimens of this species appeared in British waters during the 1960s, apparently escapes from stocking by Russia in catchments leading to the White Sea. At the time it was thought that this species might subsequently become established in the British Isles and for that reason it was included in the key by Maitland (1972a). However, although a specimen was caught in 2003, it is so rare that it is not included in the present keys.

Tilapia zillii (Gervais 1848):– formerly established in the Church Street Canal in St Helens, Lancashire, near a warm-water effluent. Now extinct with the demise of the business concerned.

Striped Grey Mullet *Mugil cephalus* Linnaeus, 1758:– a single juvenile caught in the Camel Estuary, Cornwall in October 1989 (Reay 1992). Though this was apparently just a vagrant it is likely that, as global warming proceeds, this species may start to appear more often.

In addition to the above species taken in the wild, many others are known to have been released in a variety of waters, but have never become established. These have included Fathead Minnow *Pimephales promelas* Rafinesque 1820, Brown Bullhead *Ameiurus nebulosus* (Lesueur 1819), European Mudminnow *Umbra krameri* Walbaum 1792, Danube Salmon *Hucho hucho* (Linnaeus 1758), Cutthroat Trout *Oncorhynchus clarki* (Richardson 1836), Dolly Varden

Charr *Salvelinus malma* (Walbaum 1792), American Lake Charr *Salvelinus namaycush* (Walbaum 1792) and Smallmouth Bass *Micropterus dolomieu* Lacepede 1802.

It is likely that some of these alien fish, or other temperate species, may establish at some time in the future and any strange fish should always be reported to local authorities. The problem is a world-wide one (Courtenay & Stauffer 1984; Lever 1997; IUCN 2001; Adams & Maitland 2002) and, with climate change (Maitland 1991), some southern species may now be able to thrive further north than previously. Also, some marine species, which occur too in fresh waters further south in Europe, may start to occupy suitable freshwater habitats in the British Isles at some periods of their life history. An example might be the Atlantic Silverside *Atherina presbyter* Cuvier 1829 (Palmer & Culley 1984).

NOTES ON USE OF THE KEYS

Although the keys should make it possible to name to species all fish known to occur in fresh water in Britain and Ireland, a few are rather difficult to identify without experience, and there is also the possibility of encountering hybrids (see p. 26) or a species new to the country. Since publication of the original keys (Maitland (1972a), three species new to Britain have become established, and are included in this revision. In cases of doubt, one or more of the fish concerned should be killed and preserved as described on page 12 and then sent, together with relevant details, to a competent ichthyologist for examination. It is not normally advisable to send fish by post, as they deteriorate too rapidly. When sending preserved fish, they should be drained of preservative, wrapped in damp muslin and then sealed inside a polythene bag. If this is finally placed in a box with packing and parcelled, the fish will travel for several days in perfect condition.

The keys are intended to be as useful as possible in the field so that it may be feasible to return fish to the water alive after capture, examination and identification. During this process, specimens should always be kept cool and damp. The features used to differentiate families and species are mainly external ones; characters that are as objective and absolute as possible have been selected where feasible. Nevertheless, it has been necessary in some cases to resort to characters that involve killing and dissecting the fish. The characters involved are found mainly in the head region, e.g. gills (Clupeidae), pharyngeal bones (Cyprinidae) and vomer bones (Salmonidae).

The keys follow the dichotomous pattern common to many field keys. Where possible, several distinguishing characters have been used at each point in the keys and these should always be considered in combination. Every species is illustrated and after identification using the text, the relevant text-figure should be consulted. Due consideration must always be given to the possibility of any specimen being very young, malformed, or a hybrid.

Fin rays and scales. The most common numerical features used in the keys are counts of fin rays and scales. In all fin ray counts that are mentioned, the number for each fin is obtained by counting the rays close to the body before branching occurs. Unbranched rays are included in these counts. The main scale counts are taken along the lateral line, starting at the first scale behind the operculum and ending at the last scale before the tail fin. Some diagonal scale counts are also used. These are normally counted from the lateral line up to the adipose fin (where present; see p. 11) and also from the lateral line down to the anal fin. Occasionally, counts are made from the lateral line up to the dorsal fin, and also down to the pelvic fin.

Mouth bones. The position of the main bones in the mouth is shown in Fig. 2 (p. 10). The paired premaxillary bones form the leading edge of the upper jaws, which are completed on each side by the maxillary bones. The upper jaws are opposed ventrally by lower jaws or mandibles. The single vomer bone is found centrally on the roof of the mouth and has patterns of teeth which vary according to species (Fig. 47, p. 70). The hyoid bone also occurs centrally but on the floor of the mouth and it too has patterns of teeth which vary according to species.

Pharyngeal bones. In cyprinids the pharyngeal bones (see Fig. 3, p. 15 for location) may bear one, two or three rows of teeth, depending on the species. The traditional way of denoting the dentition formula is to list the number of teeth on the left bone first, followed by a colon and then the number on the right bone. For example, when there are five teeth on the left bone and four on the right, the formula is written as 5:4. When a second (inner) row of teeth occurs, these (usually minor teeth) are separated in the formula from the major outer row by inserting a plus sign. Thus the formula for a pharyngeal bone with one minor row tooth and five major row teeth would be 5+1 (e.g. Fig. 4), and so on when three rows occur (e.g. 3+1+1 in Common Carp, for which the full dentition formula is 3+1+1:1+1+3, and 5+3+2:2+3+5 in Barbel). When one bone has more rows of teeth than the other, it is often the right. The size, firmness and spatial architecture of the bones, as well as the numbers of teeth and their shape, can be diagnostic (Fig. 18, pp. 46–47).

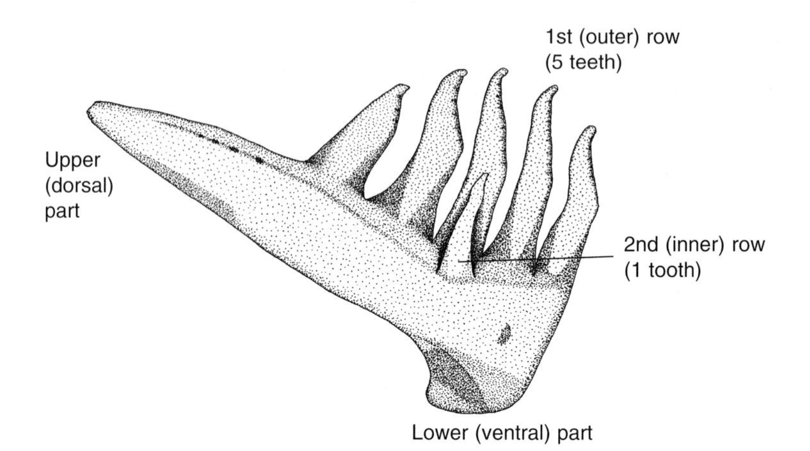

Fig. 4. The main features of a cyprinid pharyngeal bone, illustrated by the left bone of Sunbleak *Leucaspius delineatus*. The bone is orientated to show the two rows of teeth that occur in this species; the dentition formula for the bone is 5+1. A photograph of the paired bones is given by Pinder (2001, p. 129).

Body colouration. Where colours are used in the keys they refer to the condition in the fresh fish and should be true irrespective of size (above the larval and fry stages), sex and condition, unless otherwise stated. With many species it is possible to determine the sex accurately only by dissecting the gonads, especially outside the breeding season; with others, there are constant external sexual differences. These have not been included in the keys.

Hybrids

Various species of fish hybridise quite frequently in the wild. Such fish are difficult to identify, as their characteristics are usually intermediate between those of the two parent species. Unfortunately, owing to the very nature of speciation, it is often those species of fish which are most alike (and thus most difficult to identify) that are most likely to hybridise. A very wide variety of hybrids have been recorded from central Europe, but relatively few (mainly among the Cyprinidae) from the British Isles. The list of known natural hybrids recorded so far includes the following.

Alosa alosa x *Alosa fallax*
Abramis bjoerkna x *Abramis brama*
Abramis bjoerkna x *Rutilus rutilus*
Abramis bjoerkna x *Scardinius erythrophthalmus*
Abramis brama x *Leuciscus idus*
Abramis brama x *Rutilus rutilus*
Abramis brama x *Scardinius erythrophthalmus*
Alburnus alburnus x *Leuciscus cephalus*
Alburnus alburnus x *Leuciscus leuciscus*
Alburnus alburnus x *Rutilus rutilus*
Carassius auratus x *Carassius carassius*
Carassius auratus x *Cyprinus carpio*
Carassius carassius x *Cyprinus carpio*
Leuciscus cephalus x *Leuciscus leuciscus*
Leuciscus cephalus x *Rutilus rutilus*
Leuciscus leuciscus x *Scardinius erythrophthalmus*
Rutilus rutilus x *Scardinius erythrophthalmus*
Salmo trutta x *Salmo salar*
Salmo trutta x *Salvelinus fontinalis*
Platichthys flesus x *Pleuronectes platessa*

Identification of lampreys

The ammocoete larvae of lampreys are difficult to identify, whereas the adults are much easier to separate and usually distinguishable on size alone. In particular, it is always much easier to separate Sea Lamprey *Petromyzon marinus* from the two freshwater species of *Lampetra*, which closely resemble each other in the larval stages. Ivanova-Berg (1962), Maitland (1972a), Bagenal (1973) and others based their keys, which distinguished the larvae of River Lamprey *Lampetra fluviatilis* from those of Brook Lamprey *Lampetra planeri*, on the work of MacDonald (1959a,b). It has since been accepted that the latter studies are flawed and best ignored (Potter 1980; Maitland & Campbell 1992). In fact, except for the numbers of oocytes which are developed in female ammocoetes, it has so far proved impossible to separate the larvae of the two *Lampetra* species until metamorphosis, when several distinguishing characters (eye diameter, pigmentation) become apparent in the macrophthalmia (Hardisty & Potter 1971). Even the numbers of oocytes show some overlap (Hardisty 1964) and this certainly is not a character that can be used in the field.

KEY TO FAMILIES

This key separates 23 families of fish, in 14 of which only one species is known to occur in the British Isles. An outline diagram is given for a typical member of each family (Fig. 5, pp. 30–31). Separate comments on each family and keys to species are given later, as indicated by the page numbers after each family name.

1 No paired fins. Seven pairs of gill openings. No lower jaw; mouth instead a sucking disc (Fig. 5A). A single median nostril occurs between the eyes— PETROMYZONTIDAE, p. 36

— One or two pairs of fins. One pair of gill openings, each protected by an operculum. Lower jaw present; mouth never a sucking disc. Paired nostrils anterior to the eyes— **2**

2 Upper lobe of caudal fin much longer than the lower (heterocercal) (Fig. 5B). Five longitudinal rows of large bony plates on the body. Snout greatly elongated with two pairs of barbels in front of the mouth— ACIPENSERIDAE, p. 40

— Caudal fin more or less symmetrical (homocercal). No large bony plates on the body. Snout normal— **3**

3 One dorsal fin, or if two then the posterior one is small and fleshy with no rays (adipose fin) (Figs 5K–N). Pelvic fins, where present, approximately midway between the pectoral fins and the vent. Pneumatic duct (internal) present between the swimbladder and the oesophagus (physostome condition)— **4**

— Two dorsal fins (Figs 5O,P,R–V) or if one then *either* this is divided into two distinct parts, the anterior part being very spiny or replaced by isolated spines (Fig. 5Q), *or* the body is greatly flattened with both eyes on one side of the head (Fig. 5W). Pelvic fins just below or only slightly posterior to the pectoral fins. Pneumatic duct (internal) absent between the swimbladder and the oesophagus (physoclyst condition)— **15**

4 Barbels present on the head, the largest pair longer than the pectoral fins. Dorsal fin with less than 8 rays. Scales absent— **5**

— Barbels absent from the head, or if present then much shorter than the pectoral fins. Dorsal fin with more than 8 rays. Scales present (though very small in two families)— **6**

5 Six barbels, the dorsal pair very long, reaching back to the dorsal fin (Fig. 5I). Anal fin very long, with 90–92 rays— SILURIDAE, p. 66

— Eight barbels, the dorsal pair short, not reaching back beyond the operculum (Fig. 5H). Anal fin short, with only 20–22 rays— ICTALURIDAE, p. 65

6(4) Two dorsal fins, the posterior one fleshy and without rays (adipose) (Figs 5K–N)— **7**

— One dorsal fin (Figs 5C–J)— **10**

7 Scales relatively small, more than 100 along the lateral line. Red pigment often present in the skin. Pelvic axillary process present*. (Fig. 5M)— SALMONIDAE, p. 70

— Scales relatively large, less than 100 along the lateral line. Red pigment rarely present in the skin. Pelvic axillary process present* or absent— **8**

8 Lateral line present only for about the first 10 scales. Teeth well developed. Pelvic axillary process absent*. (Fig. 5K)— OSMERIDAE, p. 67

— Lateral line complete almost to the caudal fin. Teeth absent or poorly developed. Pelvic axillary process present*. (Figs 5L,N)— **9**

*The axillary process is a small dagger-shaped protuberance just above the pelvic fin.

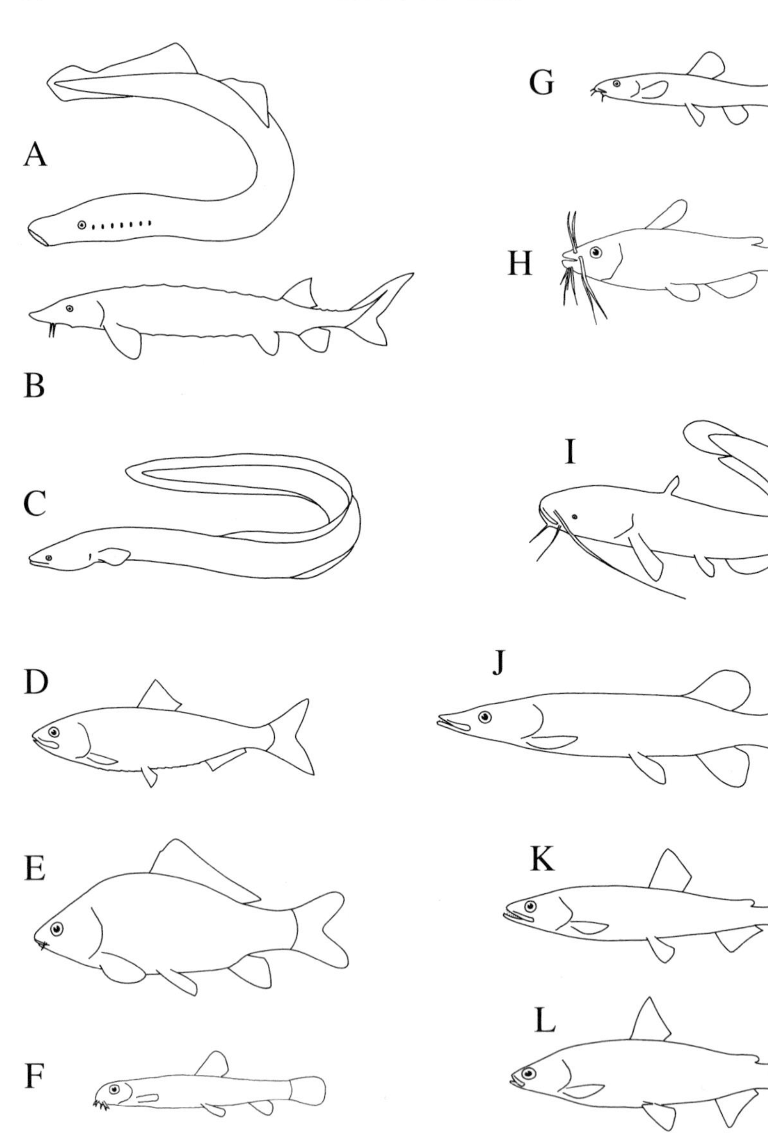

Fig. 5. Outline diagram of main family types.
A, Petromyzontidae. B, Acipenseridae. C, Anguillidae. D, Clupeidae.
E, Cyprinidae. F, Cobitidae. G, Balitoridae. H, Ictaluridae.
I, Siluridae. J, Esocidae. K, Osmeridae. L, Coregonidae.

M

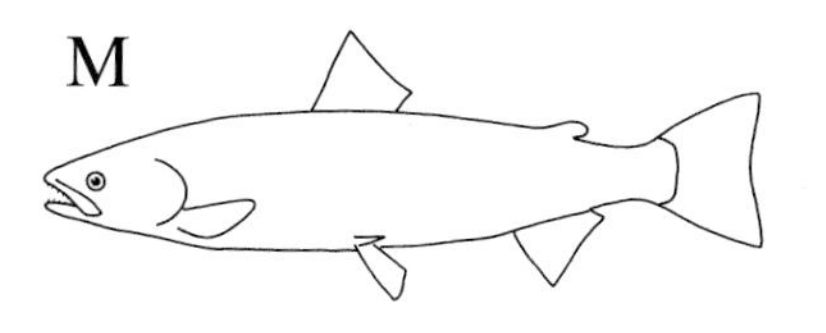

S

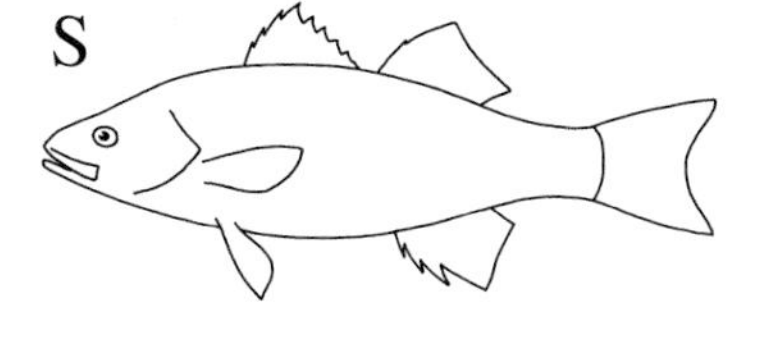

N

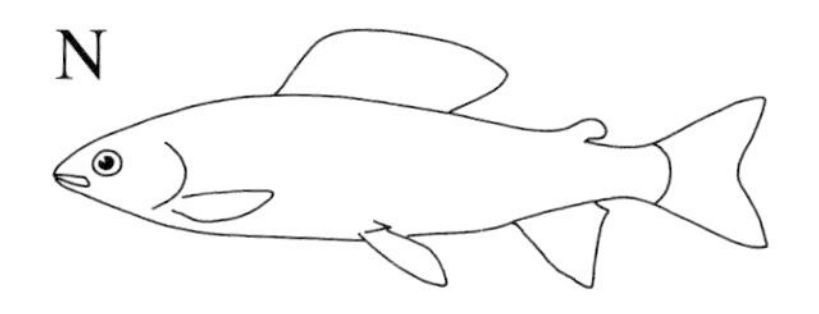

T

O

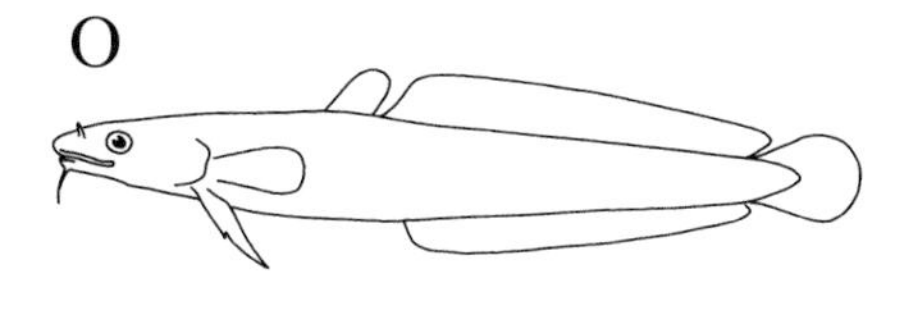

U

P

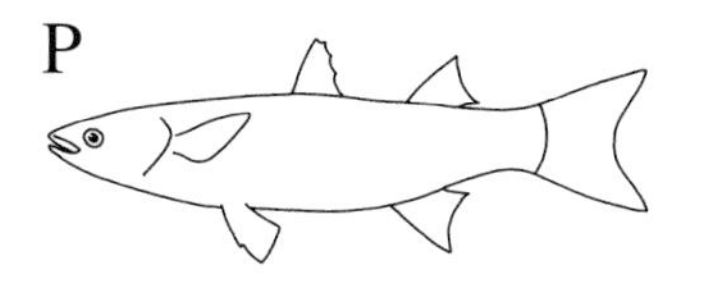

V

Q

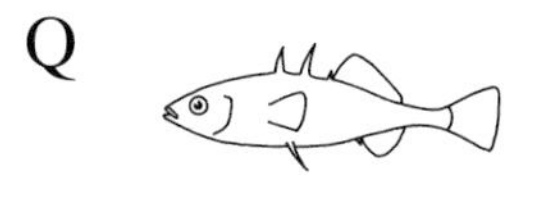

W

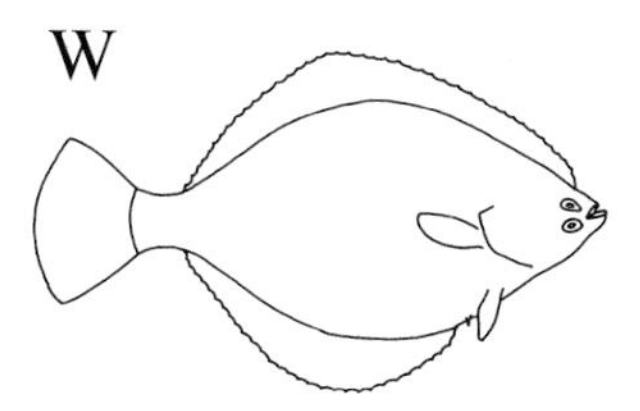

R

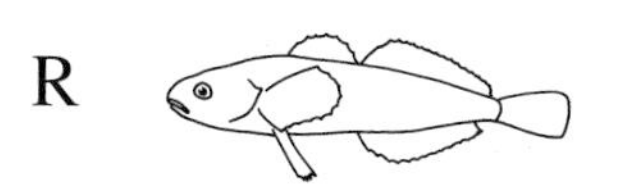

Fig. 5 (*continued*). Outline diagram of main family types. M, Salmonidae. N, Thymallidae. O, Gadidae. P, Mugilidae. Q, Gasterosteidae. R, Cottidae. S, Moronidae. T, Centrarchidae. U, Percidae. V, Gobiidae. W, Pleuronectidae.

9 Dorsal fin large with more than 20 rays, its depressed length (i.e. when pressed down against the back) much greater than that of the head (Fig. 5N). Large black pigment spots normally present in the skin. Small teeth present— THYMALLIDAE, p. 76

— Dorsal fin normally with less than 20 rays, its depressed length never greater than that of the head (Fig. 5L). Large black pigment cells never present in the skin, though small black chromatophores may be common. Teeth absent— COREGONIDAE, p. 68

10(6) Dorsal fin contiguous with caudal and anal fins. Pelvic fins absent. Gill openings reduced. Scales inconspicuous. Body extremely elongate (Fig. 5C)— ANGUILLIDAE, p. 41

— Dorsal fin distinct from caudal fin. Pelvic fins present. Gill openings normal. Scales usually obvious. Body not extremely elongate— **11**

11 Scales on the ventral surface keeled*. Lateral line absent. Obvious adipose eyelids. Large elongate scales over the inner part of the caudal fin. (Fig. 5D)— CLUPEIDAE, p. 42

— Scales on the ventral surface not keeled*. Lateral line present. No obvious eyelids. No elongate scales over the inner part of the caudal fin— **12**

*Ventrally, along the mid-line in shad, there is a sharp median ridge, often developed into a spine posteriorly.

12 Head elongate; mouth very large with well-developed teeth. Dorsal fin mostly posterior to the vent (Fig. 5J). Scales present on the head— ESOCIDAE, p. 66

— Head normal; mouth moderate or small with teeth absent or poorly developed. Dorsal fin entirely or mostly anterior to the vent. Scales absent from the head— **13**

13 Less than five barbels on the head (Fig. 5E). Mouth normal. Scales on the body usually distinct— CYPRINIDAE, p. 44

— More than five barbels on the head. Mouth small. Scales on the body indistinct— **14**

14 Bifid spine in a pocket under each eye. All barbels the same length (Fig. 5F)— COBITIDAE, p. 64

— No bifid spine under the eye. One pair of barbels shorter than the others (Fig. 5G)— BALITORIDAE, p. 64

15(3) Two dorsal fins, or if one it is divided into two distinct parts, the anterior part being very spiny. Body never greatly flattened or with isolated spines— **16**

— One dorsal fin. Body greatly flattened or with a row of dorsal spines— **22**

16 Head with a single barbel below the mouth and one small barbel beside each nostril (Fig. 5O). Anal fin with more than 60 rays— GADIDAE, p. 77

— Head without barbels. Anal fin with less than 60 rays— **17**

17 Scales absent over most of the body. Anterior dorsal fin rays flexible. (Fig. 5R)— COTTIDAE, p. 81

— Well developed scales over most of the body. Anterior dorsal fin rays rigid— **18**

18 Lateral line absent. Less than 5 spiny rays in the anterior dorsal fin. Dorsal fins widely separated, the distance between them always exceeding the length of the longest dorsal ray (Fig. 5P)— MUGILIDAE, p. 78

— Lateral line present. More than 5 spiny rays in the anterior dorsal fin. Dorsal fins continuous or close together, the distance between them never exceeding the length of the longest dorsal ray— **19**

19 Two or fewer anal fin spines present. Either less than or more than 9 or 10 spiny rays in the first dorsal fin— **20**

— Three or more anal fin spines present. 9 or 10 spiny rays in the first dorsal fin— **21**

20 Two anal fin spines present. Tail fin concave posteriorly (Fig. 5U). More than 10 spiny rays in the first dorsal fin. Pelvic fins not joined medially— PERCIDAE, p. 84

— Anal fin spines absent. Tail fin convex posteriorly (Fig. 5V). Less than 9 spiny rays in the first dorsal fin. Pelvic fins joined medially— GOBIIDAE, p. 86

21(19) Second dorsal fin with 3 spiny rays anteriorly. More than 70 scales along the lateral line. Anal fin concave (Fig. 5S)—MORONIDAE, p. 81

— Second dorsal fin with 1 spiny ray anteriorly. Less than 70 scales along the lateral line. Anal fin convex (Fig. 5T)— CENTRARCHIDAE, p. 82

22(15) Three or more strong spines anterior to the dorsal fin. Body not flattened, with eyes on either side of the head. Pelvic fins with less than three rays. Each pelvic fin represented by a strong spine. (Fig. 5Q)—
GASTEROSTEIDAE, p. 80

— No spines anterior to the dorsal fin. Body extremely flattened (Fig. 5W), with both eyes on one side of the head (usually the right). Pelvic fins with more than 3 rays. Pelvic fins without spines—
PLEURONECTIDAE, p. 87

KEYS TO SPECIES

Key to Lamprey Family PETROMYZONTIDAE

1 Teeth absent, sucker incomplete, oral hood over the mouth. Eyes indistinct and opaque (Fig. 6). Branchial groove present. Skin semi-transparent, dull on the underside— Ammocoete larvae, **2**

— Sucker complete. Eyes visible— **4**

2 Dark pigment cells over most of the oral hood and base of the caudal fin (Fig. 6A). More than 68 trunk myomeres (muscle bands)— **Sea Lamprey *Petromyzon marinus***

— No distinct pigmentation on ventral area of the oral hood or on the caudal fin (Fig. 6B). Less than 68 trunk myomeres— *LAMPETRA*, **3**

[NOTE: It is not possible to distinguish the larvae of the two species of *Lampetra* in the field (see p. 27). Larvae more than 12 cm in length are likely to be *Lampetra planeri*. With dead specimens in the laboratory, estimates of the numbers of oocytes in females can be used as indicated below in couplet 3.]

3 Mean number of oocytes in transverse sections typically greater than 50— **River Lamprey *Lampetra fluviatilis***

— Mean number of oocytes in transverse sections typically less than 50— **Brook Lamprey *Lampetra planeri***

4(1) Height of 2nd dorsal fin approximately the same as body height at that point— Transformers, **5**

— Height of 2nd dorsal fin higher than body height at that point— Adults, **7**

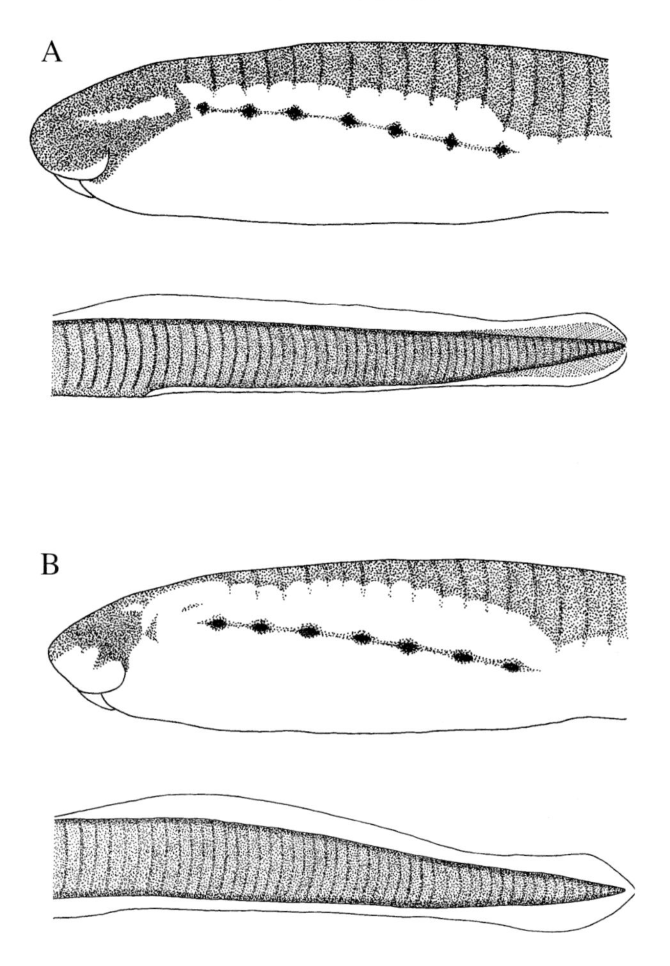

Fig. 6. Ammocoete larvae of Petromyzontidae.
Head regions above and tail regions below for: A, *Petromyzon*; B, *Lampetra*.

5 Teeth on the oral disc numerous and in radiating rows (Fig. 7A). More than 68 trunk myomeres— **Sea Lamprey *Petromyzon marinus***

— Teeth on the oral disc few and not in radiating rows (Figs 7B,C). Less than 68 trunk myomeres— *LAMPETRA*, **6**

6 Eyes large, body silvery, snout long. Teeth often sharp. Usually less than 12 cm in length— **River Lamprey *Lampetra fluviatilis***

— Eyes small, body less silvery, snout short. Teeth blunt. Often, but not always, more than 12 cm in length—
Brook Lamprey *Lampetra planeri*

7(4) Teeth on the oral disc close together in radiating rows (Fig. 7A); anterior dental plate with two large teeth. Back and sides of the body with a marbled pattern. Length at spawning usually 45–90 cm—
Sea Lamprey *Petromyzon marinus* (Fig. 8; Plates 3, 4; p. 135)

— Teeth on the oral disc widely spaced and not in radiating rows (Figs 7B,C); anterior dental plate with, at most, one small tooth. Back and sides of the body uniform in colour. Length at spawning less than 45 cm— *LAMPETRA*, **8**

8 Posterior lamina with 7–10 cusps (arrow, Fig. 7B). Most teeth strong and sharp. Dorsal fins usually separate, though they may be joined in sexually mature fish. Length at spawning usually 25–40 cm and almost always more than 17 cm—
River Lamprey *Lampetra fluviatilis* (Fig. 9; Plates 1, 4; p. 133)

— Posterior lamina with 5–9 cusps (arrow, Fig. 7C). All teeth weak and blunt. Dorsal fins normally connected or almost so. Length at spawning usually 10–15 cm and almost always less than 17 cm—
Brook Lamprey *Lampetra planeri* (Fig. 10; Plate 2; p. 134)

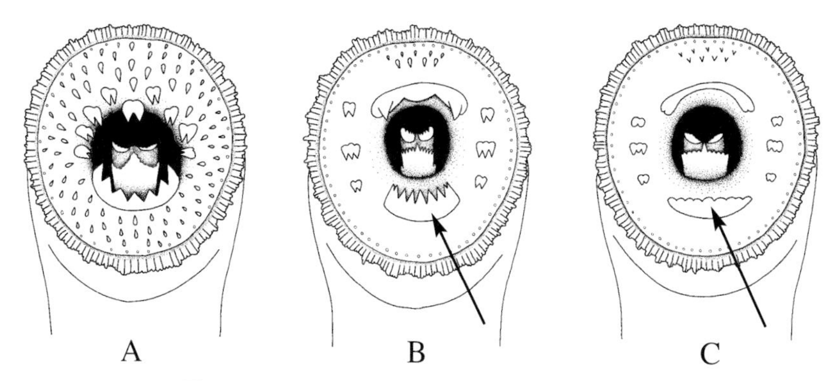

Fig. 7. Mouth structure of adult Petromyzontidae.
A, Sea lamprey *Petromyzon marinus*. B, River Lamprey *Lampetra fluviatilis*.
C, Brook Lamprey *Lampetra planeri*.

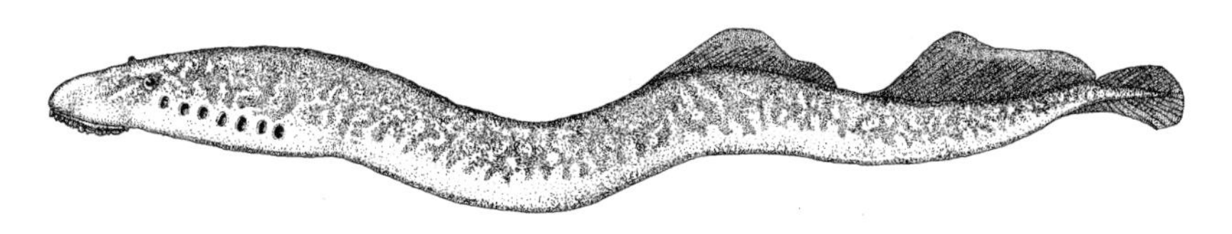

Fig. 8. Sea Lamprey *Petromyzon marinus*.

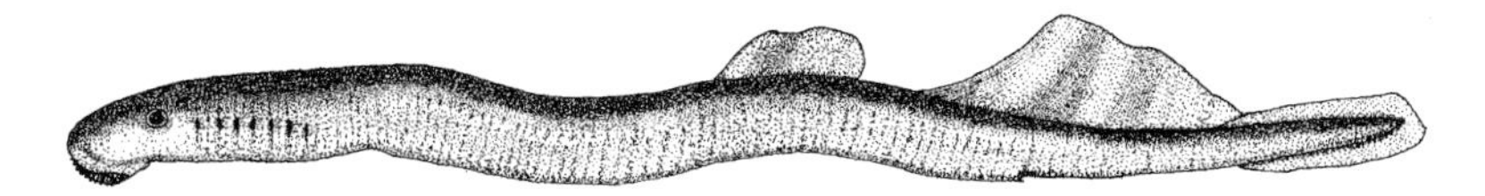

Fig. 9. River Lamprey *Lampetra fluviatilis*.

Fig. 10. Brook Lamprey *Lampetra planeri*.

Key to Sturgeon Family ACIPENSERIDAE

Only one native species has occurred in the British Isles: the Common Sturgeon *Acipenser sturio* (Fig. 11). This is very rare now and the occasional specimens that are caught could be confused with either the Siberian Sturgeon *Acipenser baeri* Brandt 1869, which is now farmed in several countries in Europe – with some escapes to coastal waters around Britain – or the Sterlet *Acipenser ruthenus* Linnaeus 1758, which is widely sold in Britain as an aquarium or pond fish, occasional specimens turning up in British waters. The three species may be separated by the key below.

1 Lateral scutes more than 50. Barbels with fimbria (delicate fringing processes). Gill rakers not fan-shaped— **Sterlet *Acipenser ruthenus***

— Lateral scutes less than 50. Barbels without fimbria— **2**

2 Gill rakers fan-shaped. Lateral scutes 42–47— **Siberian Sturgeon *Acipenser baeri***

— Gill rakers not fan-shaped. Lateral scutes 24–36— **Common Sturgeon *Acipenser sturio*** (Fig. 11; p. 136)

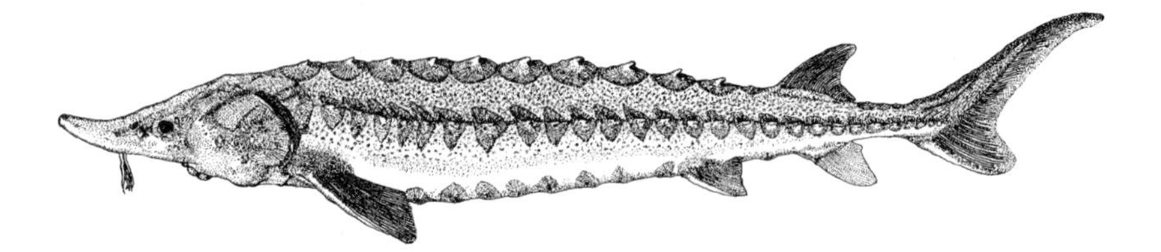

Fig. 11. Common Sturgeon *Acipenser sturio*.

Eel Family ANGUILLIDAE

One species occurs in Britain and Ireland—
European Eel ***Anguilla anguilla*** (Fig. 12; p. 137)

Fig. 12. Line-drawing and photograph of European Eel *Anguilla anguilla*.

Key to Herring Family CLUPEIDAE

1 Main longitudinal row of lateral scales numbering more than 70. More than 60 gill rakers on the first gill arch (Fig. 13A)—
Allis Shad ***Alosa alosa*** (Fig. 14; Plate 5; p. 138)

— Main longitudinal row of lateral scales numbering less than 70. Less than 60 gill rakers on the first gill arch (Fig. 13B)—
Twaite Shad ***Alosa fallax*** (Fig. 15; Plates 5, 6; p. 139)

Note: hybrids occur between these two species and show intermediate characters.

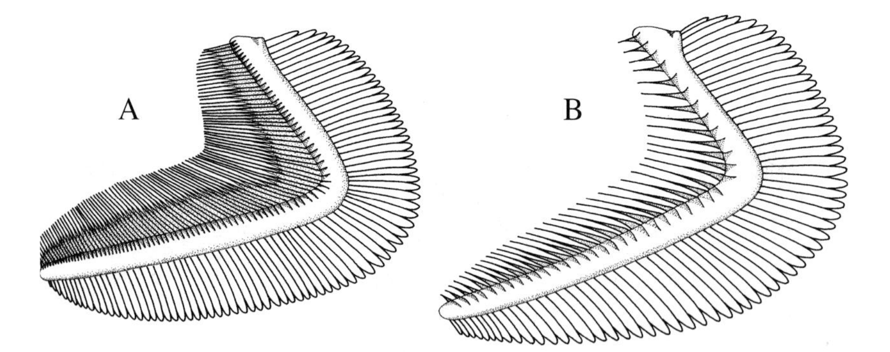

Fig. 13. Gills of Clupeidae.
A, Allis Shad *Alosa alosa*. B, Twaite Shad *Alosa fallax*.

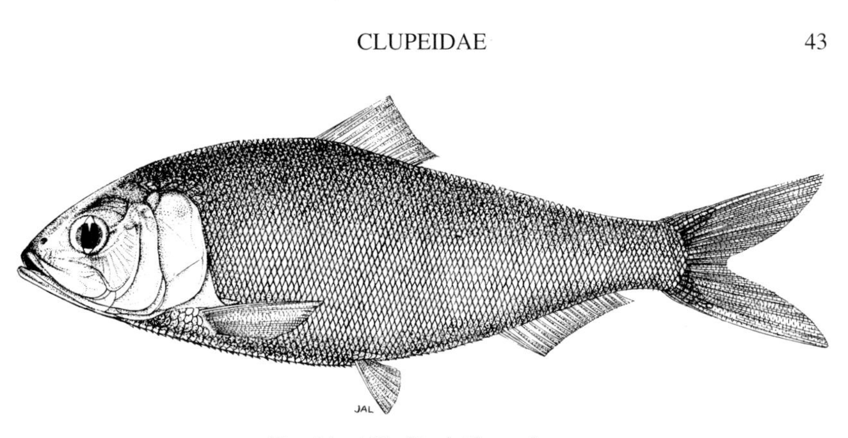

Fig. 14. Allis Shad *Alosa alosa*.

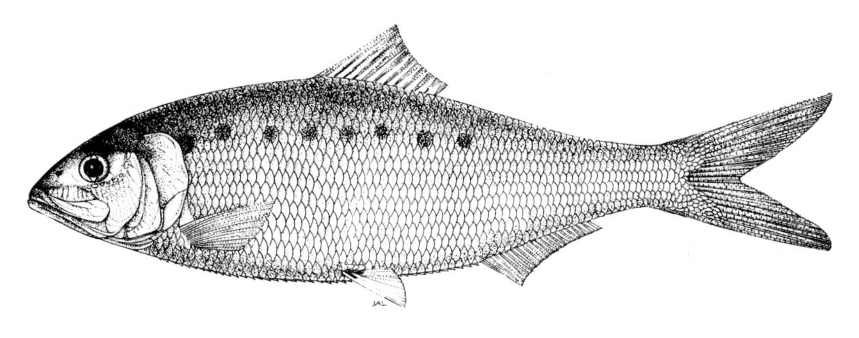

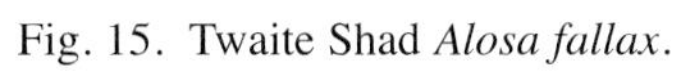
Fig. 15. Twaite Shad *Alosa fallax*.

Key to Carp Family CYPRINIDAE

Two new species, the Sunbleak *Leucaspius delineatus* and the False Harlequin *Pseudorasbora parva*, have become established in England since the first edition of this key was published (Maitland 1972a). In addition, the Grass Carp *Ctenopharyngodon idella* (Valenciennes 1844) – though unable to breed here – is commonly used for weed control in standing waters. All three species are included in the key below.

1 Barbels present on the head— **2**

— Barbels absent from the head— **5**

2 Dorsal fin with more than 15 rays. Less than 40 scales along the lateral line*. Pharyngeal teeth 3+1+1:1+1+3 (Fig. 18, p. 46)—
Common Carp *Cyprinus carpio* (Fig. 16; Plate 17; p. 144)

*Note that in some strains the covering of scales is much reduced (Mirror Carp) or even absent (Leather Carp).

— Dorsal fin with less than 15 rays. More than 40 scales along the lateral line. Pharyngeal teeth not as above— **3**

3 More than 90 scales along the lateral line. Dorsal fin convex. Pharyngeal teeth in one row of 4 or 5 (Fig. 18)—
Tench *Tinca tinca* (Fig. 17; Plate 18; p. 151)

— Less than 60 scales along the lateral line. Dorsal fin concave. Pharyngeal teeth in two or three rows— **4**

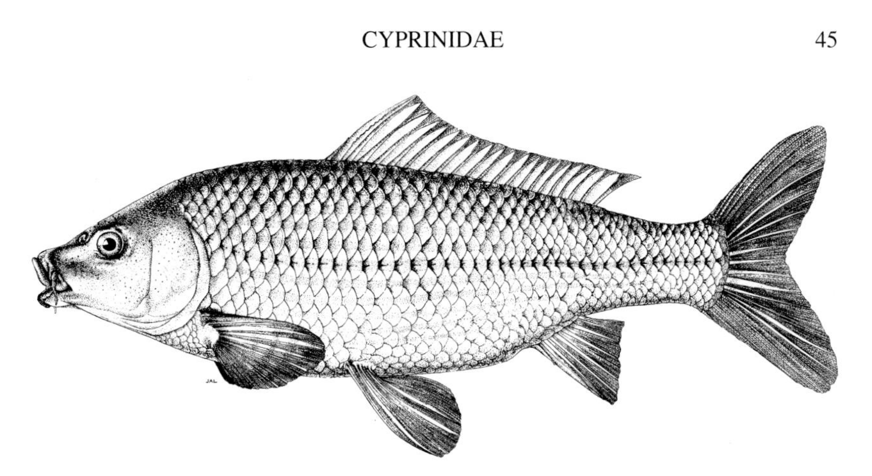

Fig. 16. Common Carp *Cyprinus carpio*

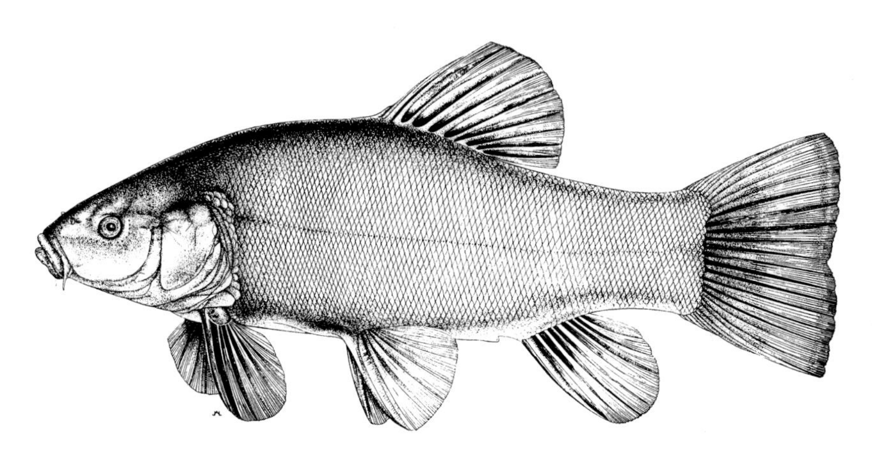

Fig. 17. Tench *Tinca tinca*.

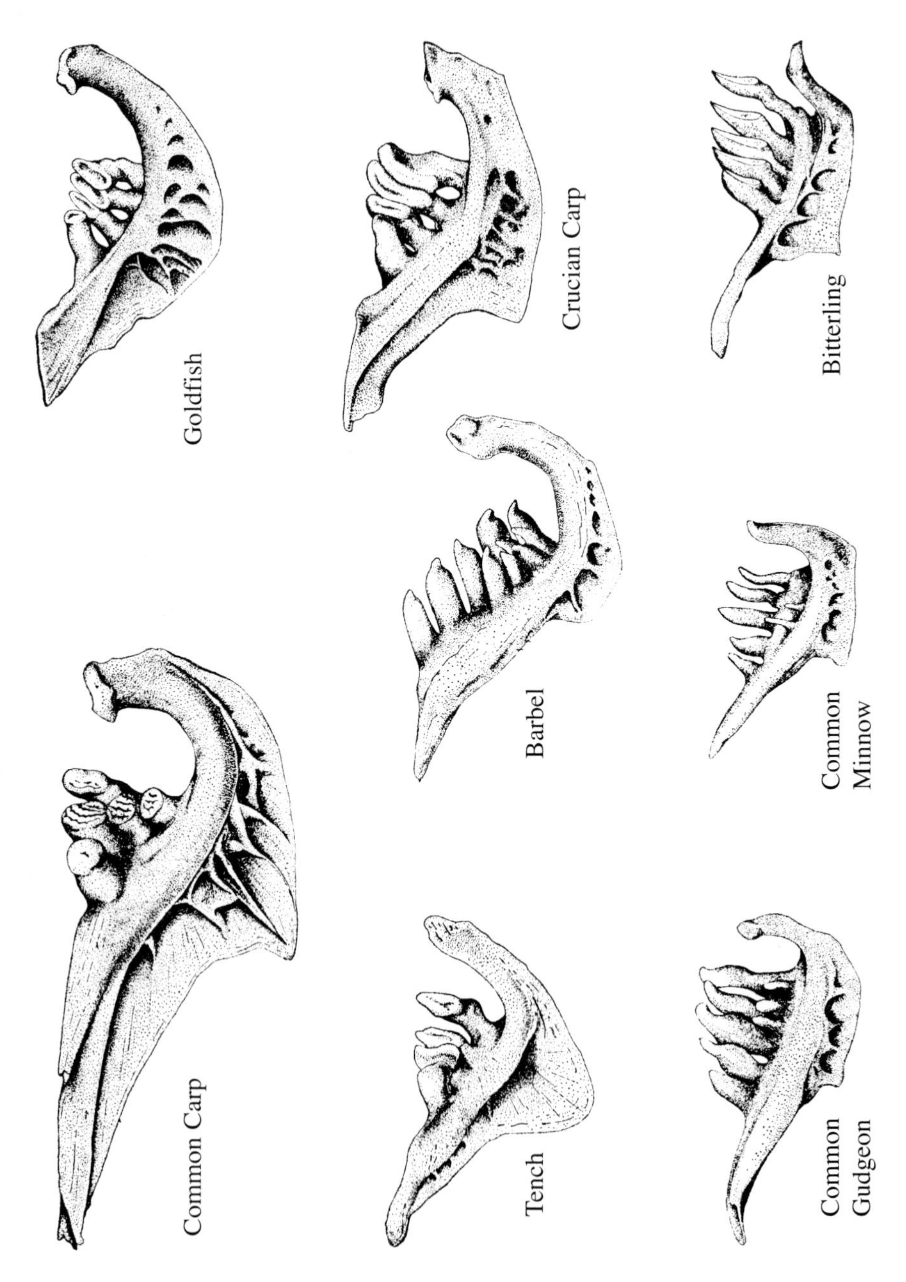
Goldfish
Crucian Carp
Bitterling
Barbel
Common Minnow
Common Carp
Tench
Common Gudgeon

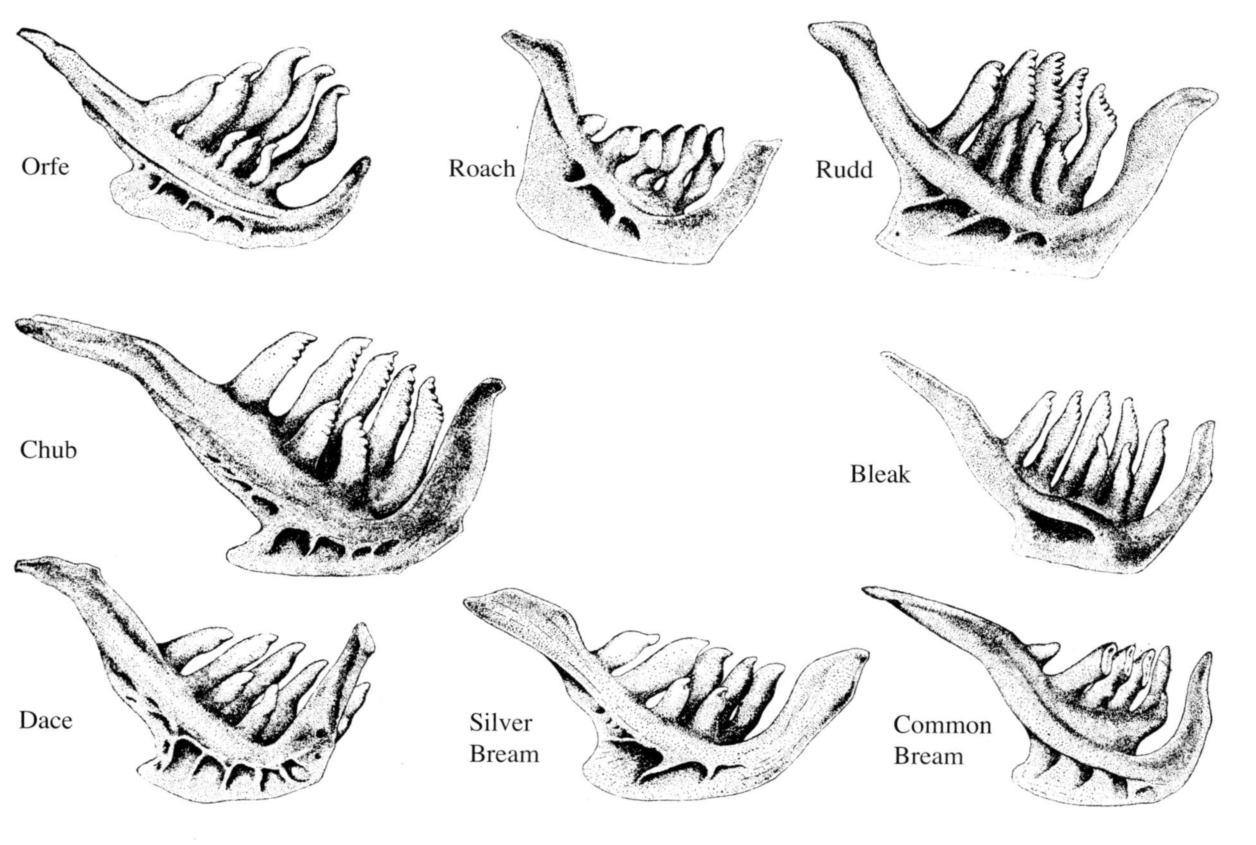

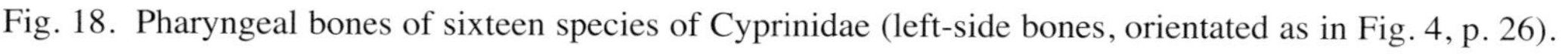
Fig. 18. Pharyngeal bones of sixteen species of Cyprinidae (left-side bones, orientated as in Fig. 4, p. 26).

4 Four barbels on the head. More than 50 scales along the lateral line. Pharyngeal teeth in three rows, 5+3+2:2+3+5 (Fig. 18)—
Barbel *Barbus barbus* (Fig. 19; Plate 15; p. 142)

— Two barbels on the head Less than 50 scales along the lateral line. Pharyngeal teeth in two rows, 5+3:2/3+5 (Fig. 18)—
Common Gudgeon *Gobio gobio* (Fig. 20; Plate 16; p. 145)

5(1) Less than 15 branched rays in the anal fin— **6**

— More than 15 branched rays in the anal fin— **16**

6 More than 14 rays in the dorsal fin. Less than 35 scales along the lateral line. Pharyngeal teeth 4:4— **7**

— Less than 12 rays in the dorsal fin. More than 35 scales along the lateral line. Pharyngeal teeth other than 4:4— **8**

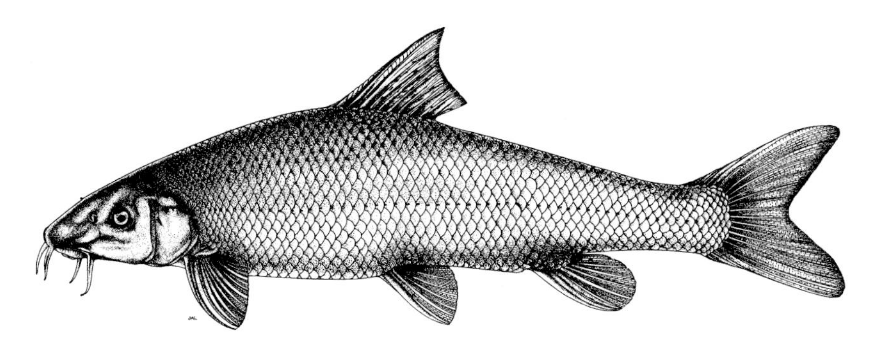

Fig. 19. Barbel *Barbus barbus*.

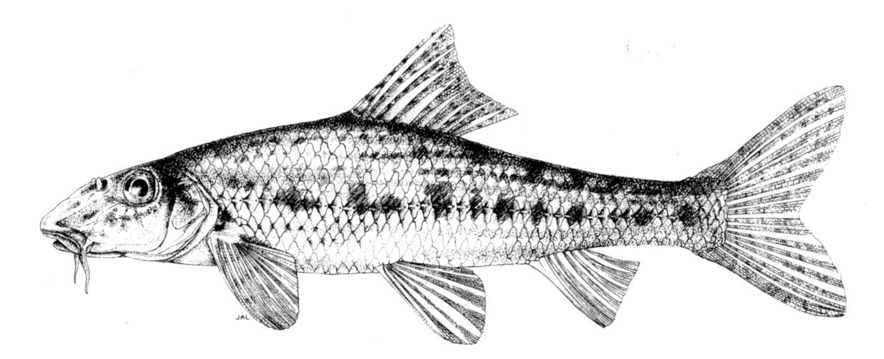

Fig. 20. Common Gudgeon *Gobio gobio*.

7 Less than 31 scales along the lateral line. More than 34 gill rakers on the first gill arch. First dorsal ray strong, coarsely serrated. Dorsal fin concave. Pharyngeal teeth 4:4 (Fig. 18)—
Goldfish *Carassius auratus* (Fig. 21; p. 143)

— More than 31 scales along the lateral line. Less than 34 gill rakers on the first gill arch. First dorsal ray feeble, weakly serrated. Dorsal fin convex. Pharyngeal teeth 4:4 (Fig. 18)—
Crucian Carp *Carassius carassius* (Fig. 22; Plate 8; p. 144)

8(6) Eyes small and on the side of the head near the midline. Eye diameter less than 40% of distance between the snout and the front of the eye—
Grass Carp *Ctenopharyngodon idella*

— Eyes large and on the side of the head above the midline. Eye diameter more than 40% of distance between the snout and the front of the eye— **9**

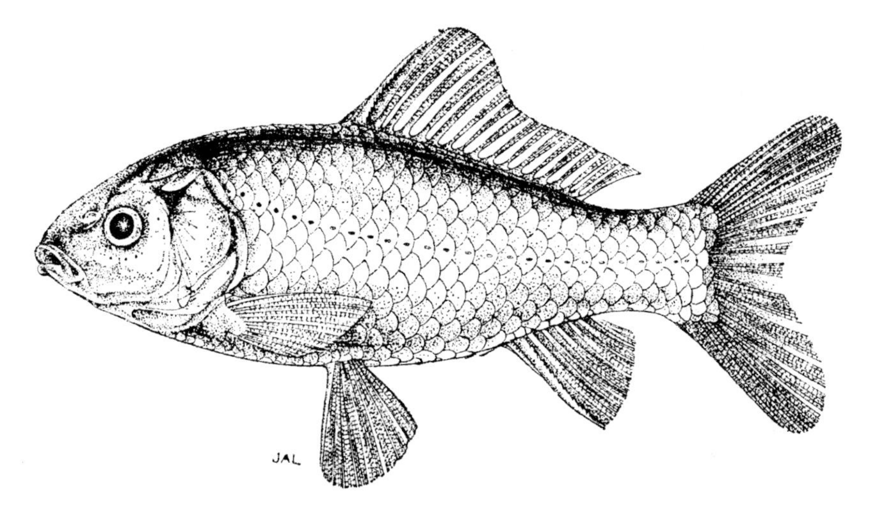

Fig. 21. Goldfish *Carassius auratus*.

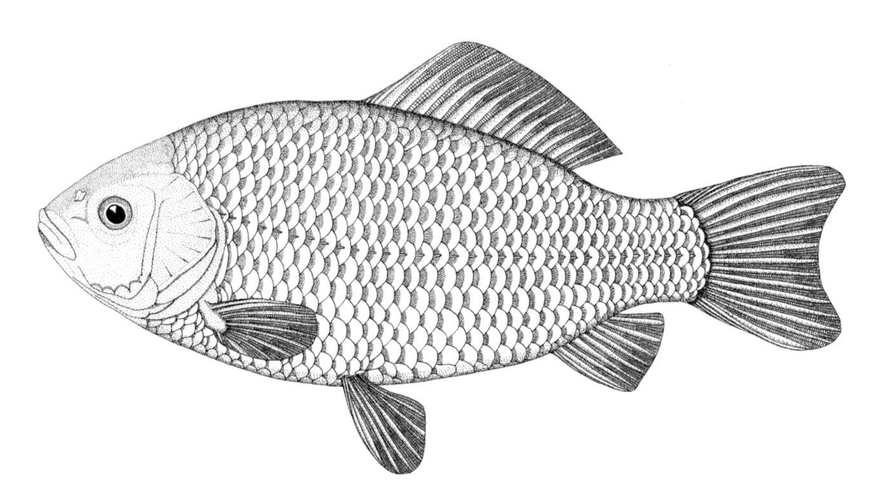

Fig. 22. Crucian carp *Cyprinus carassius*.

9 Less than 39 scales along the lateral line. Less than 14 rays in the pectoral fin— **10**

— More than 39 scales along the lateral line. More than 14 rays in the pectoral fin— **11**

10 Lateral line normal. A dark band of equal width runs along the sides. Anal fin with 6 branched rays. Pharyngeal teeth 5:5—
False Harlequin *Pseudorasbora parva* (Fig. 23; Plate 13; p. 149)

— Lateral line short, confined to first 4–7 scales. A bluish stripe along the sides, which widens on the caudal peduncle. Anal fin with 8–10 branched rays. Pharyngeal teeth 5:5 (Fig. 18)—
Bitterling *Rhodeus sericeus* (Fig. 24; Plate 14; p. 149)

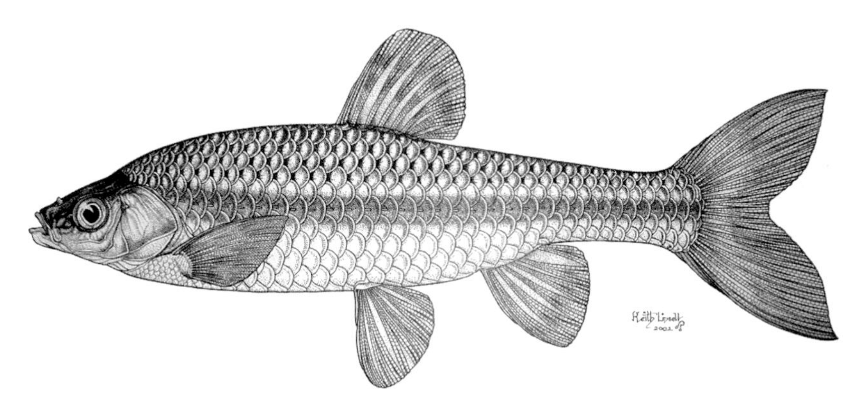

Fig. 23. False Harlequin *Pseudorasbora parva*.

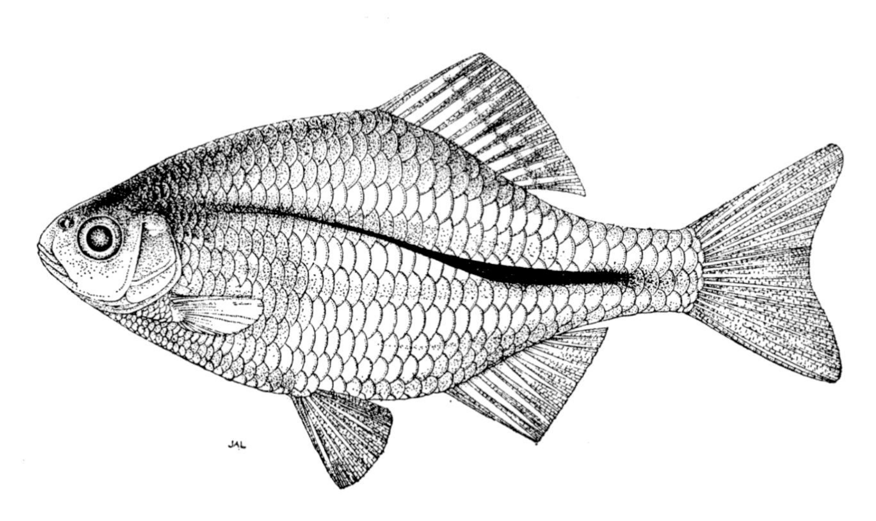

Fig. 24. Bitterling *Rhodeus sericeus*.

11(9) More than 80 scales along the lateral line, which is usually incomplete behind the middle of the body (Fig. 25A). Tubules of the lateral line extending to the free edge of the scales. Pharyngeal teeth 5+2:2+4/5 (Fig. 18)—

Common Minnow *Phoxinus phoxinus* (Fig. 26; Plate 11; p. 148)

— Less than 60 scales along the lateral line, which is always complete (Fig. 25B). Tubules of the lateral line not extending to the free edge of the scales— **12**

12 More than 54 scales along the lateral line; 9–10 scales between the dorsal fin and the lateral line; 5–6 scales between the anal fin and the lateral line. Pharyngeal teeth 5+3:3+5—

Orfe *Leuciscus idus* (Fig. 27; Plate 21; p. 147)

— Less than 54 scales along the lateral line; 7–8 scales between the dorsal fin and the lateral line; 3–4 scales between the anal fin and the lateral line— **13**

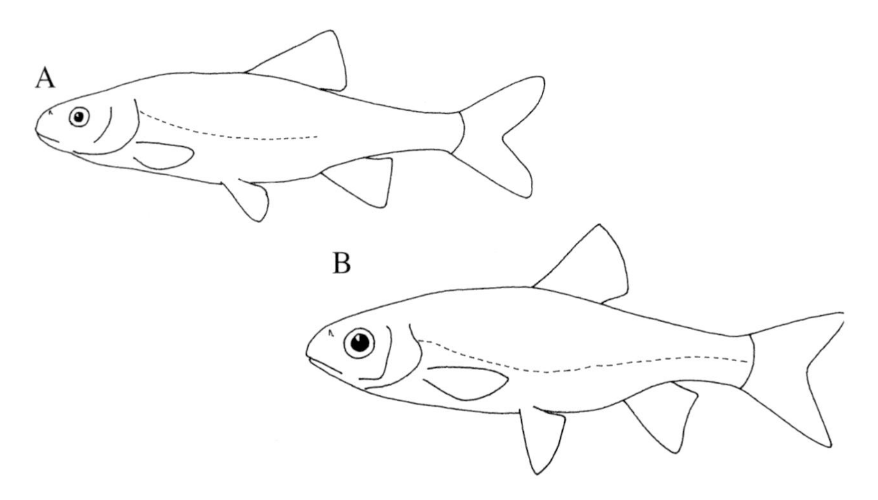

Fig. 25. Diagrams illustrating the extent of the lateral lines in A, Common Minnow and B, Orfe.

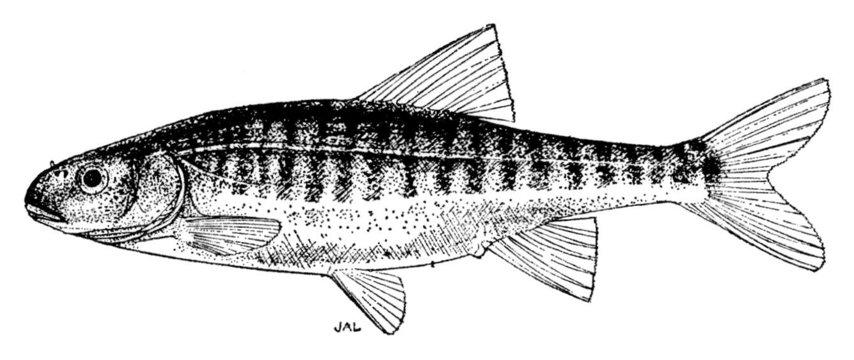

Fig. 26. Common Minnow *Phoxinus phoxinus*.

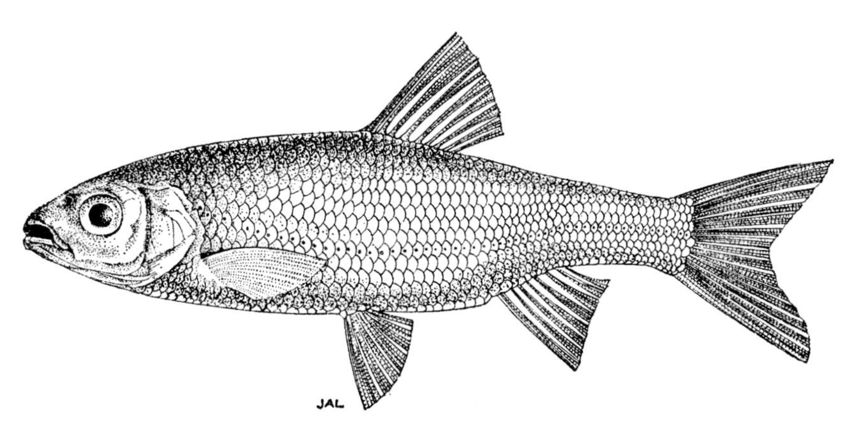

Fig. 27. Orfe *Leuciscus idus*.

13 Less than 13 (usually 10–12) rays in the anal fin. Length of the body more than 3.5 times its maximum depth. 44–53 scales along the lateral line. Pharyngeal teeth 5+2:2/3+5— **14**

— More than 11 (usually 13–14) rays in the anal fin. Length of the body less than 3.5 times its maximum depth. 40–50 scales along the lateral line. Pharyngeal teeth not as above— **15**

14 Less than 47 scales along the lateral line. Dorsal fin originating behind the base of the pelvic fin. Anal and dorsal fins straight or slightly convex. Fork to the caudal fin shallow. Pharyngeal teeth 5+2:2+5 (Fig. 18)—
Chub *Leuciscus cephalus* (Fig. 28; Plate 20; p. 146)

— More than 46 scales along the lateral line. Dorsal fin originating above the base of the pelvic fin. Anal and dorsal fins concave. Fork to the caudal fin deep. Pharyngeal teeth 5+2:2/3+5 (Fig. 18)—
Dace *Leuciscus leuciscus* (Fig. 29; Plate 22; p. 147)

Fig. 28. Chub *Leuciscus cephalus*.

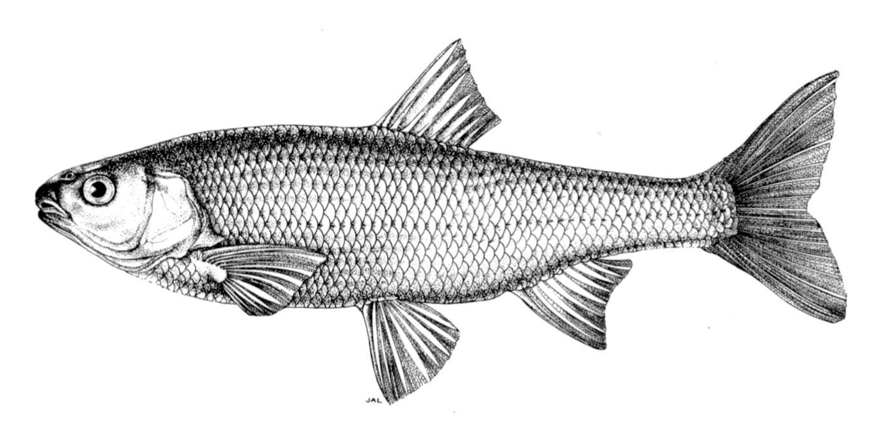

Fig. 29. Dace *Leuciscus leuciscus*.

15(13) Scales along the lateral line 42–45. Front of the dorsal fin above or very slightly behind the base of the pelvic fins. One row of pharyngeal teeth (5:5 or 6:6) on which the pectination (tooth-like ridges at the ends of the teeth) is weak or absent (Fig. 18, p. 47)—

Roach ***Rutilus rutilus*** (Fig. 30; Plate 9; p. 150)

— Scales along the lateral line 40–43. Front of the dorsal fin distinctly behind the base of the pelvic fins. Two rows of pharyngeal teeth (5+3:3+5) on which the pectination (tooth-like ridges at the ends of the teeth) is strong (Fig. 18)—

Rudd ***Scardinius erythrophthalmus*** (Fig. 31; Plate 10; p. 151)

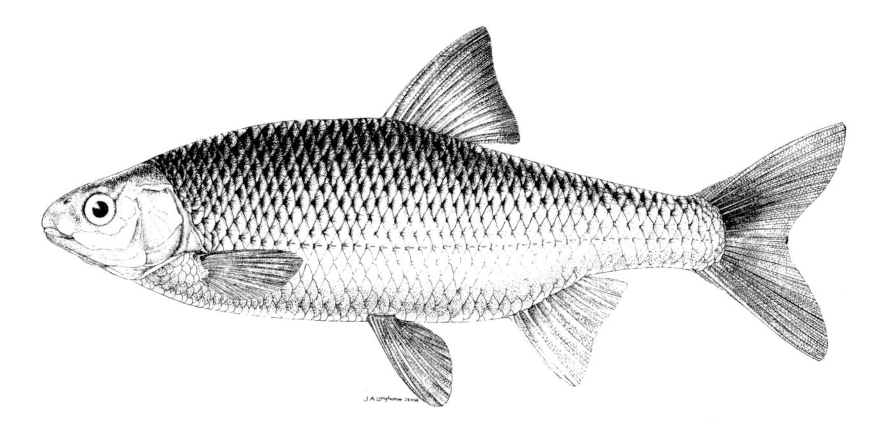

Fig. 30. Roach *Rutilus rutilus*.

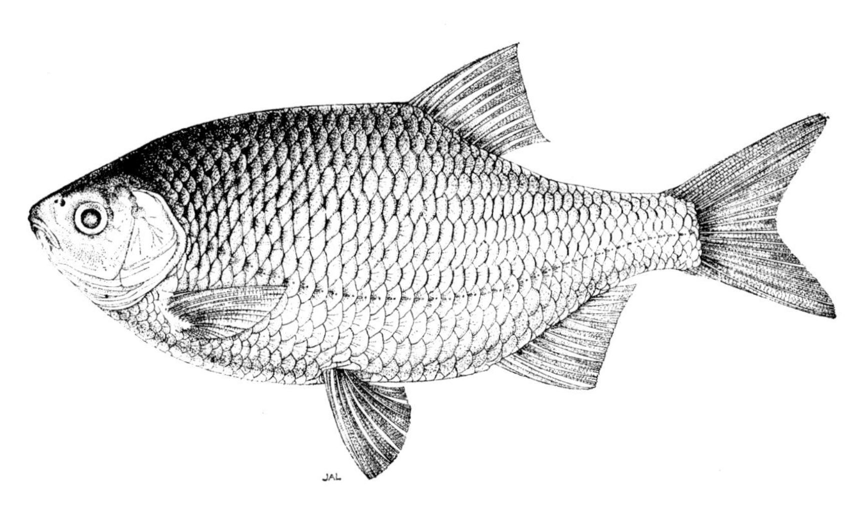

Fig. 31. Rudd *Scardinius erythrophthalmus*.

16(5) Length of body more than four times its maximum depth. Less than 21 rays in the anal fin. Less than 5 scales between the anal fin and the lateral line— **17**

— Length of body less than three times its maximum depth. More than 20 rays in the anal fin. More than 4 scales between the anal fin and the lateral line— **18**

17 Lateral line short, only up to the 10th scale; lateral line dark. Anal fin with 15–17 branched rays. Pharyngeal teeth 5+1:0/1+4 (Fig. 4, p. 26)—
Sunbleak *Leucaspius delineatus* (Fig. 32; Plate 12; p. 146)

— Lateral line normal, with no dark pigment. Anal fin with 17–20 branched rays. Pharyngeal teeth 5+2:2+5 (Fig. 18, p. 47)—
Bleak *Alburnus alburnus* (Fig. 33; Plate 19; p. 142)

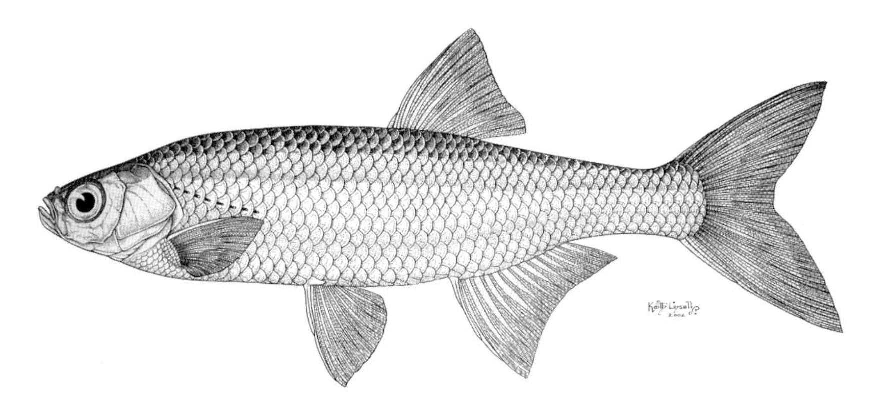

Fig. 32. Sunbleak *Leucaspius delineatus*.

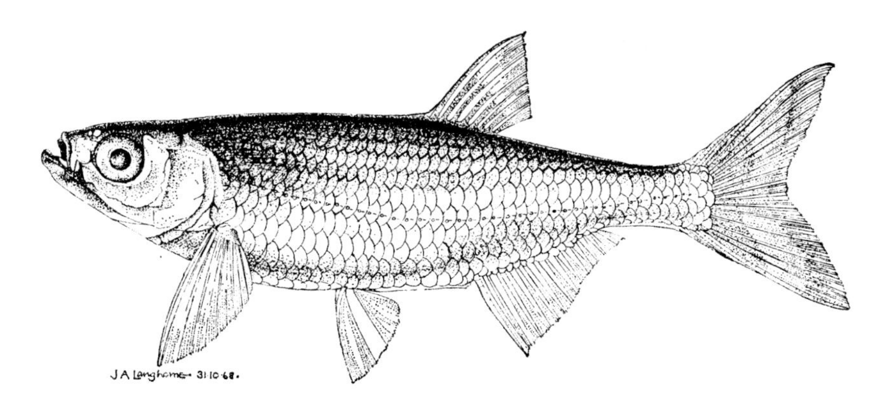

Fig. 33. Bleak *Alburnus alburnus*.

18(16) Less than 50 scales along the lateral line. Less than 27 rays in the anal fin. 8–11 scales between the dorsal fin and the lateral line. Distance between the tip of the snout and the eye usually less than or equal to the diameter of the eye. Two rows of pharyngeal teeth, 5+2:2+5 (Fig. 18, p. 47)— **Silver Bream** ***Abramis bjoerkna*** (Fig. 34; Plate 7; p. 140)

— More than 49 scales along the lateral line. More than 25 rays in the anal fin. 11–15 scales between the dorsal fin and the lateral line. Distance between the tip of the snout and the eye usually more than the diameter of the eye. One row of pharyngeal teeth, 5:5 (Fig. 18)—
Common Bream ***Abramis brama*** (Fig. 35; p. 141)

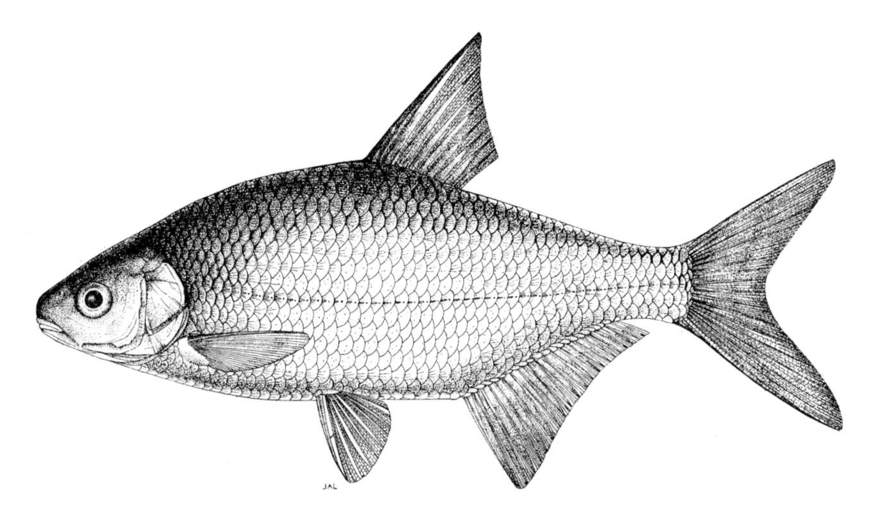

Fig. 34. Silver Bream *Abramis bjoerkna*.

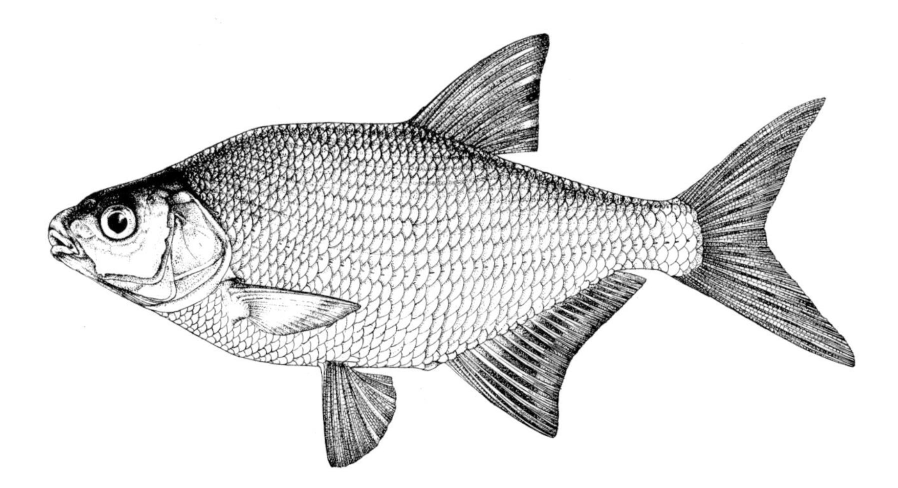

Fig. 35. Common Bream *Abramis brama*.

Spined Loach Family COBITIDAE

One species occurs in Britain (not found in Ireland). Six short barbels (Fig. 36A)— **Spined Loach *Cobitis taenia*** (Fig. 37; Plate 23; p. 152)

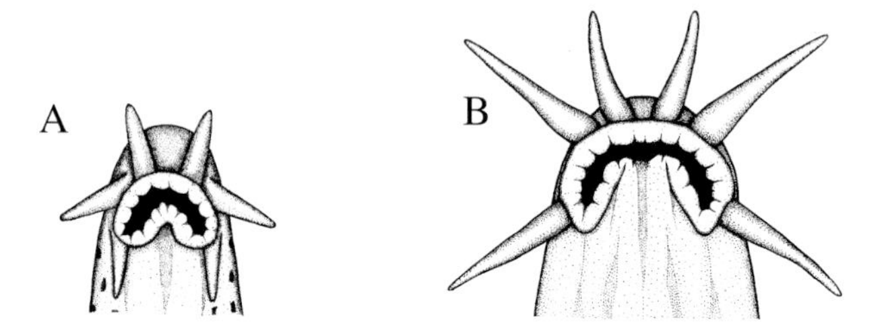

Fig. 36. Mouth structure of A, Cobitidae: Spined Loach *Cobitis taenia*, and B, Balitoridae: Stone Loach *Barbatula barbatula*.

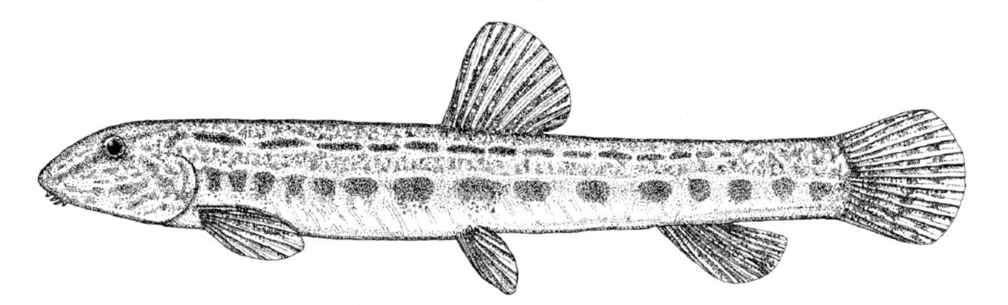

Fig. 37. Spined Loach *Cobitis taenia*.

Stone Loach Family BALITORIDAE

One species occurs in Britain and Ireland. Six relatively long barbels (Fig. 36B)— **Stone Loach *Barbatula barbatula*** (Fig. 38; Plate 24; p. 153)

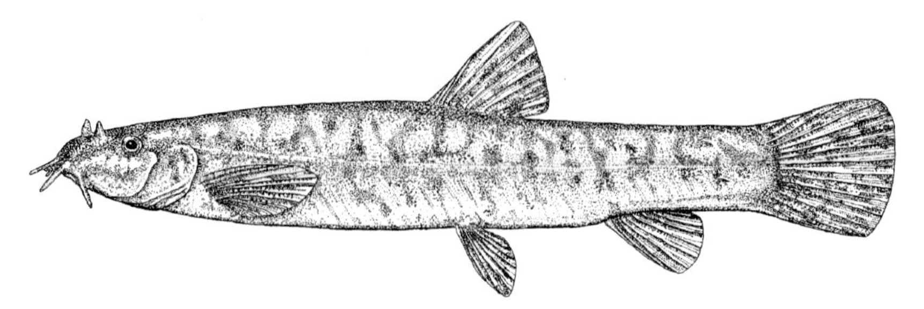

Fig. 38. Stone Loach *Barbatula barbatula*.

Key to American Catfish Family ICTALURIDAE

One species is established in Britain: the Black Bullhead *Ameiurus melas* (Fig. 39). However, two other North American catfish may occur, the Channel Catfish *Ictalurus punctatus* (Rafinesque 1818) and the Brown Bullhead *Ameiurus nebulosus* (Lesueur 1819), for both are available in aquarium shops and could escape from garden ponds or be released intentionally. Both are included in the key below. All are scaleless, with eight barbels around the mouth, of which the dorsal pair are relatively short. There is an adipose fin and the pectoral fins are armed with spines.

1 Caudal fin deeply forked—
Channel Catfish *Ictalurus punctatus* (Plate 25)

— Caudal fin emarginate, not deeply forked— **2**

2 Barbs on posterior pectoral spines weak or absent. 17–21 anal fin rays. 17–19 gill rakers. Dorsal fin membranes darkened—
Black Bullhead *Ameiurus melas* (Fig. 39; Plate 25; p. 153)

— Barbs on posterior pectoral spines large. 21–24 anal fin rays. 13–15 gill rakers. Dorsal fin membranes not darkened—
Brown Bullhead *Ameiurus nebulosus*

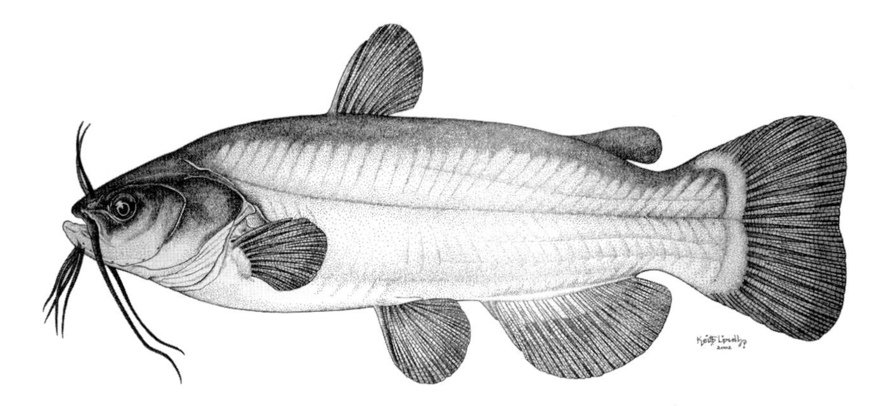

Fig. 39. Black Bullhead *Ameiurus melas*.

European Catfish Family SILURIDAE

One species occurs in Britain (not in Ireland)—
Wels Catfish *Silurus glanis* (Fig. 40; Plate 26; p. 154)

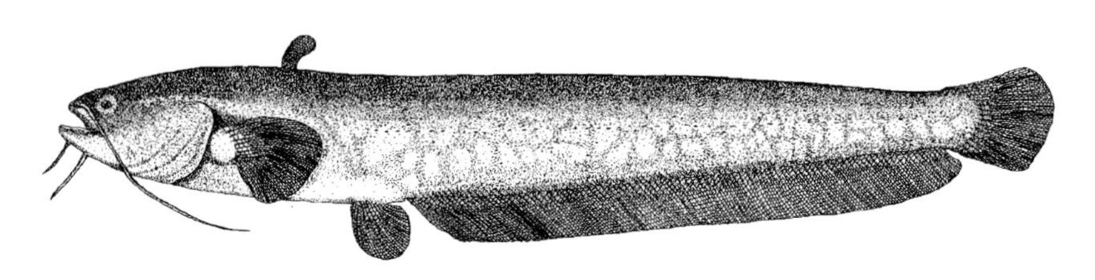

Fig. 40. Wels Catfish *Silurus glanis*.

Pike Family ESOCIDAE

One species occurs in Britain and Ireland—
Pike *Esox lucius* (Fig. 41; Plate 36; p. 155)

Fig. 41. Pike *Esox lucius*.

Smelt Family OSMERIDAE

One species occurs in Britain and Ireland—
Smelt *Osmerus eperlanus* (Fig. 42; p. 156)

Fig. 42. Line-drawing and photograph of Smelt *Osmerus eperlanus*.

Key to Whitefish Family COREGONIDAE

1 Snout conical and produced in front. Distance between the anterior edge of the eye and the tip of the snout more than twice the diameter of the eye— **Houting *Coregonus oxyrinchus*** (Fig. 43; p. 159)

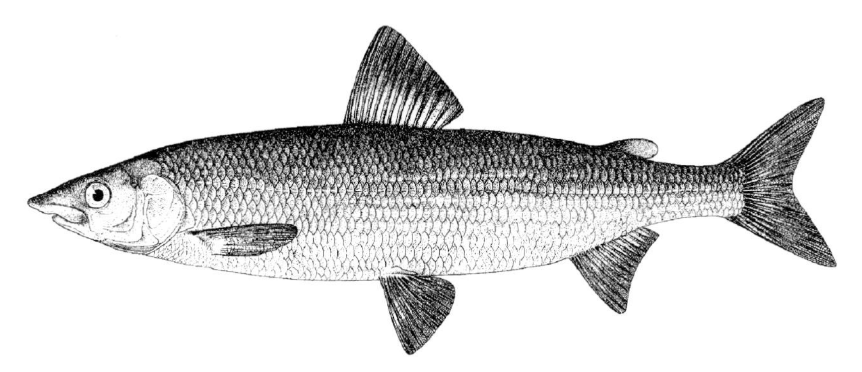

Fig. 43. Houting *Coregonus oxyrinchus*.

— Snout not produced in front. Distance between the anterior edge of the eye and the tip of the snout less than twice the diameter of the eye— **2**

2 Mouth terminal. Found only in Ireland— **Pollan *Coregonus autumnalis*** (Fig. 44; Plate 27; p. 157)

— Mouth other than terminal. Found only in Britain— **3**

3 Mouth superior. Less than 13 rays in the anal fin— **Vendace *Coregonus albula*** (Fig. 45; Plate 29; p. 157)

— Mouth inferior. 13 or more rays in the anal fin— **Powan *Coregonus lavaretus*** (Fig. 46; Plate 30; p. 158)

Figs 44–46 *(On page 69)*: Coregonidae. Fig. 44: Pollan *Coregonus autumnalis*. Fig. 45: Vendace *Coregonus albula*. Fig. 46: Powan *Coregonus lavaretus*.

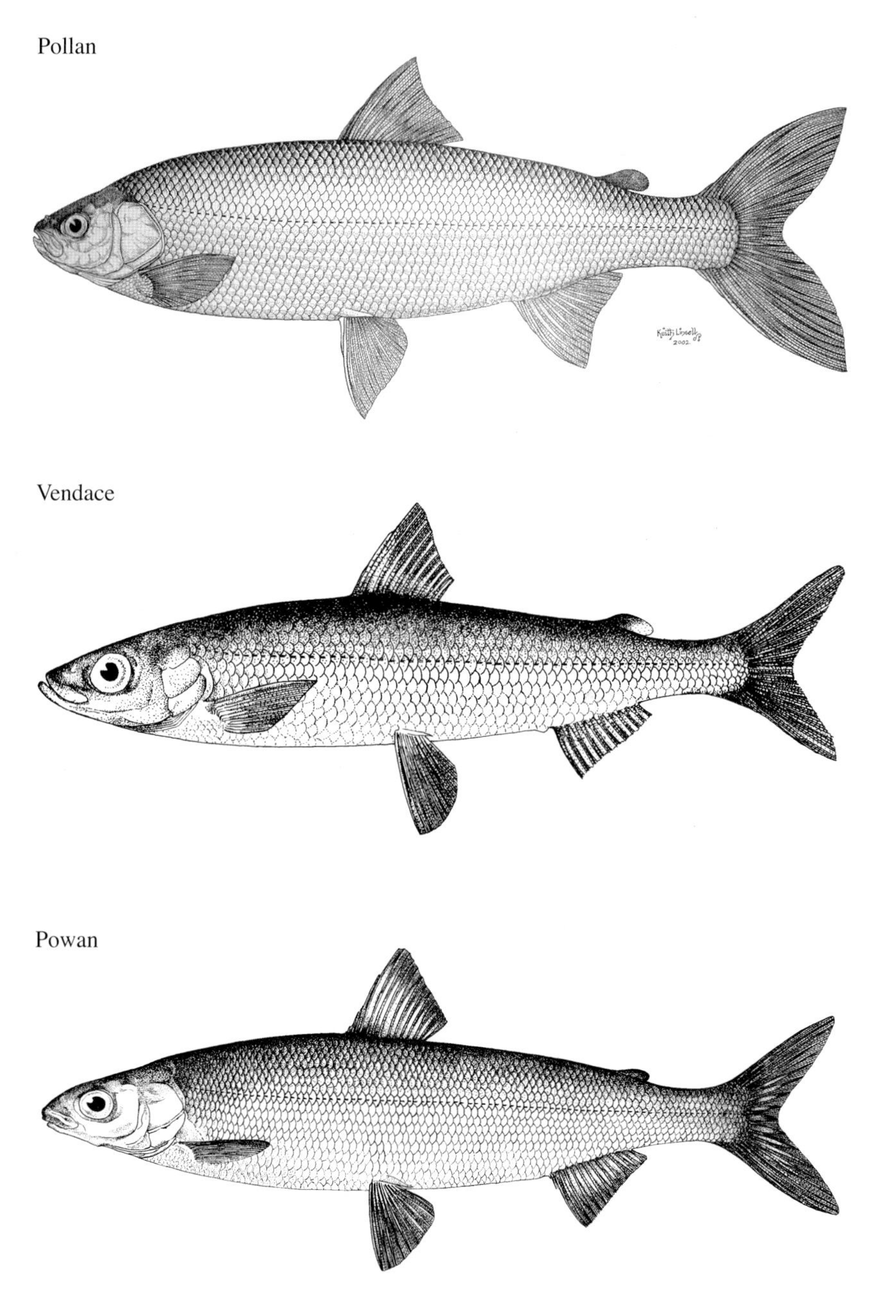
Pollan
Vendace
Powan

Key to Salmon Family SALMONIDAE

1 More than 160 scales along the lateral line. Vomerine teeth confined to the head of the vomer (Fig. 47), whose shaft is toothless (Figs 47D,E)— **2**

— Less than 150 scales along the lateral line. Vomerine teeth not confined to the head of the vomer, whose shaft has many teeth (Figs 47A–C)— **3**

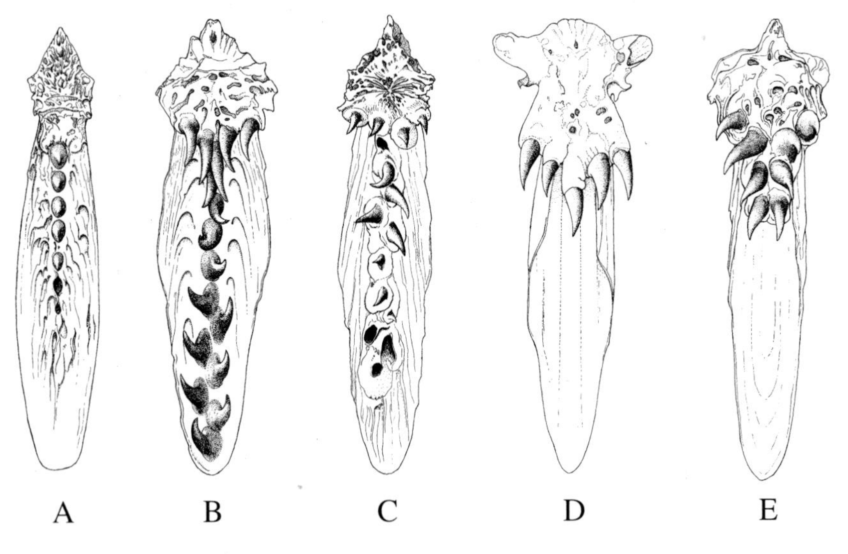

Fig. 47. Vomer bones of Salmonidae.
A, Atlantic Salmon *Salmo salar;*
B, Brown Trout *Salmo trutta*; C, Rainbow Trout *Oncorhynchus mykiss*;
D, Arctic Charr *Salvelinus alpinus*; E, Brook Charr *Salvelinus fontinalis*.

2 Hyoid teeth (see Fig. 2, p. 10) present. Back mainly uniformly coloured. Red spots on sides without blue borders. No dark wavy lines on dorsal or caudal fins. No black stripe on the anal fin—

Arctic Charr *Salvelinus alpinus* (Fig. 48; Plate 33; p. 163)

Fig. 48. Arctic Charr *Salvelinus alpinus*.

— Hyoid teeth absent. Back strongly vermiculated. Red spots on sides often with blue borders. Dark wavy lines on dorsal and caudal fins. Black stripe present on the anal fin—

Brook Charr *Salvelinus fontinalis* (Fig. 49; Plate 34; p. 164)

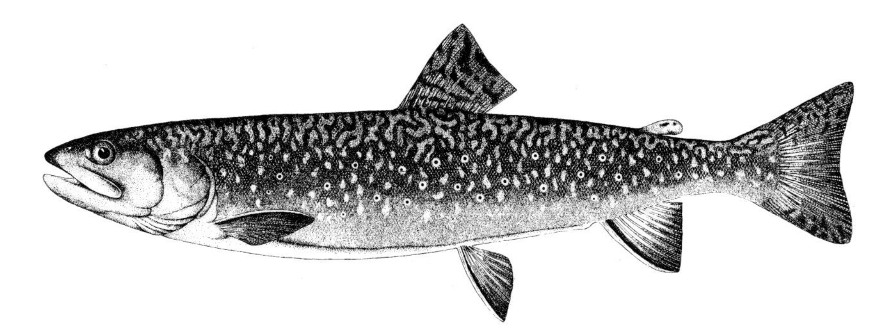

Fig. 49. Brook Charr *Salvelinus fontinalis*.

3(1) More than 130 scales along the lateral line. No red spots on the body, but a broad pink band is normally present along either side. Numerous black spots on the body and fins, especially the adipose and tail fins—
Rainbow Trout *Oncorhynchus mykiss* (Fig. 50; Plate 31; p. 160)

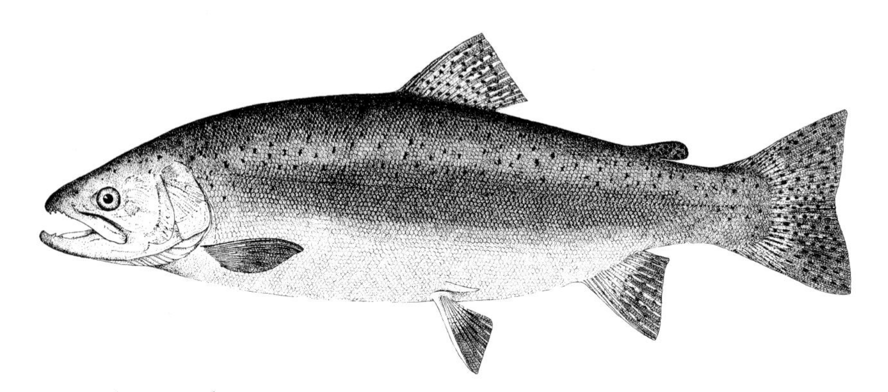

Fig. 50. Rainbow Trout *Oncorhynchus mykiss*.

— Less than 130 scales along the lateral line. The body may be completely silver, but normally has many black and some red spots present; never a broad band along either side. Black spots on the adipose and caudal fins ill-defined or absent— **4**

4 Length usually less than 15 cm. Parr marks (a single line of dark lateral patches, Fig. 51) usually present, or body completely silvery. Tail fin distinctly forked— Juveniles, **5**

— Length usually more than 15 cm. Parr marks usually absent (though the body may be well spotted). Tail fin indistinctly or not forked— Adults, **8**

5 Parr marks more or less distinct along the body; numerous spots present, mostly black but some red—
Fry and parr (the latter occuring all year round), **6**

— Parr marks indistinct or absent; body almost completely silvery, though a few black spots may be present— Smolts, **7**

6 Parr marks 10–12. A few faint black spots on the dorsal fin which has 10–12 rays. Operculum with less than 3 spots. Adipose fin normally brown. Caudal peduncle thin; tail fin with a deep fork and pointed ends. Pectoral fins large, when stretched back often reaching behind the level of the origin of the dorsal fin. Maxilla reaching to about the middle of the eye— **Parr stage of Atlantic Salmon *Salmo salar*** (Fig. 51A)

— Parr marks 9–10. Many definite black spots on the dorsal fin which has 8–10 rays. Operculum with more than 3 black spots. Adipose fin normally red. Caudal peduncle thick; tail fin with a shallow fork and rounded ends. Pectoral fins normal, when stretched back not reaching behind the level of the origin of the dorsal fin. Maxilla reaching to midway between the pupil and the rear of the eye— **Parr stage of Brown Trout *Salmo trutta*** (Fig. 51B)

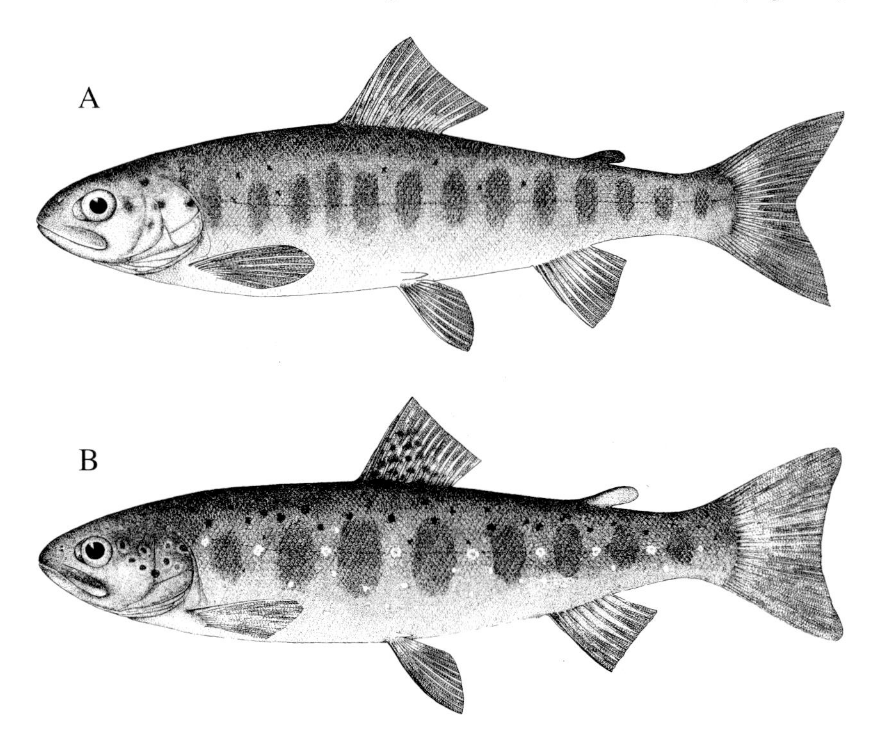

Fig. 51. Immature stage (parr) of *Salmo*.
A, Atlantic Salmon *Salmo salar*. B, Brown Trout *Salmo trutta*.

7(5) Dorsal fin with 10–12 rays. 10–13 scales between the adipose fin and the lateral line. Operculum with less than 3 spots. Caudal peduncle thin; tail fin with a deep fork. Pectoral fins large—

Smolt stage of Atlantic Salmon *Salmo salar*

— Dorsal fin with 8–10 rays. 13–16 scales between the adipose fin and the lateral line. Operculum with more than 3 spots. Caudal peduncle thick; tail fin with a shallow fork. Pectoral fins normal—

Smolt stage of Brown Trout *Salmo trutta*

8(4) Head of vomer toothless, the shaft poorly toothed with deciduous teeth (Fig. 47A). Dorsal fin with 10–12 rays. 10–13 scales between the adipose fin and the lateral line. When laid back, the last ray of the anal fin usually extends about as far back as the first ray—

Adult stage of Atlantic Salmon *Salmo salar* (Fig. 52; p. 161)

— Head of vomer toothed; the shaft also well toothed with persistent teeth (Fig. 47B). Dorsal fin with 8–10 rays. 13–16 scales between the adipose fin and the lateral line. When laid back, the last ray of the anal fin does not extend as far posteriorly as the first ray—

Adult stage of Brown Trout *Salmo trutta* (Fig. 53; Plate 32; p. 162)

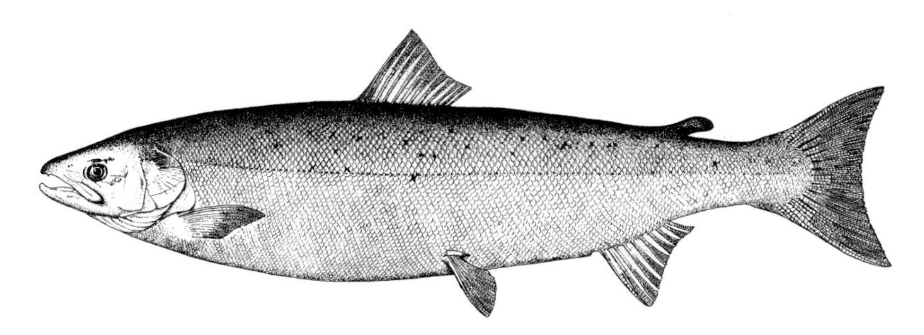

Fig. 52. Atlantic Salmon *Salmo salar*.

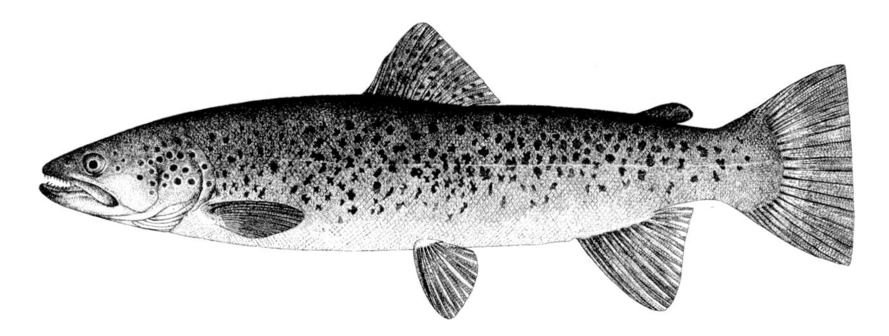

Fig. 53. Brown Trout *Salmo trutta*.

Grayling Family THYMALLIDAE

One species occurs in Britain (not found in Ireland)—
Grayling *Thymallus thymallus* (Fig. 54; Plate 28; p. 164).

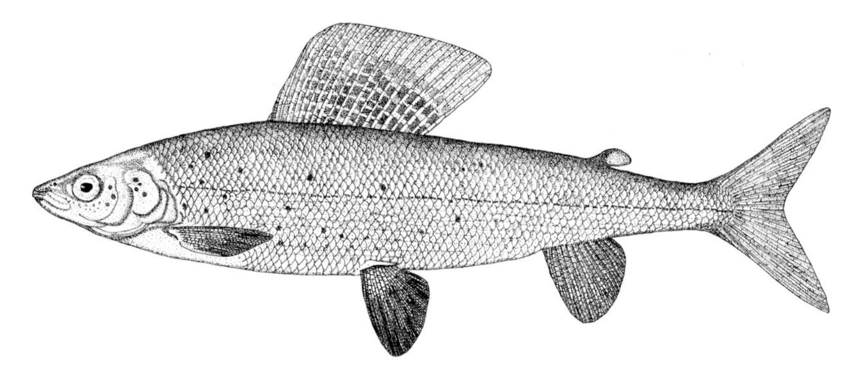

Fig. 54. Grayling *Thymallus thymallus*.

Cod Family GADIDAE

One species occurs in Britain (not found in Ireland)—
Burbot *Lota lota* (Fig. 55; Plate 35; p. 165).

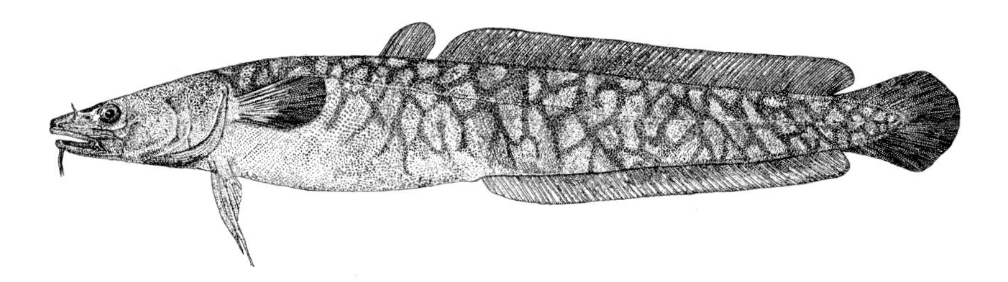

Fig. 55. Burbot *Lota lota*.

Key to Grey Mullet Family MUGILIDAE

1 Upper lip thick, its depth greater than one-tenth of the head length and more than half the diameter of the eye. No scales present on the lower jaw—
Thick-lipped Grey Mullet *Chelon labrosus* (Fig. 56; Plate 45; p. 166)

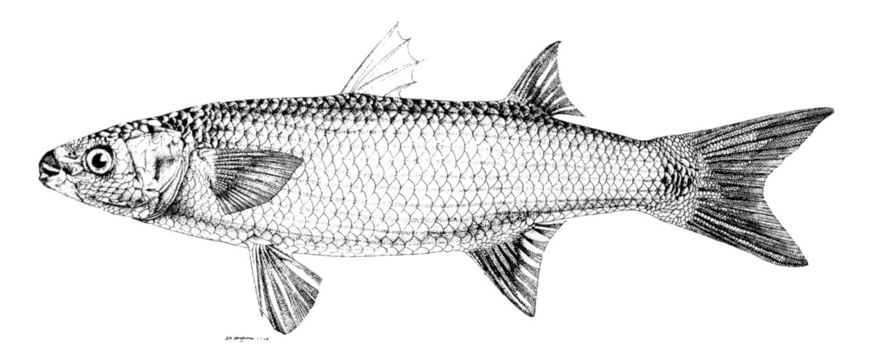

Fig. 56. Thick-lipped Grey Mullet *Chelon labrosus*.

— Upper lip thinner, its depth less than one-tenth of the head length and less than half the diameter of the eye. Scales present on the lower jaw— **2**

2 Scales on the dorsal side of the head not extending to the nostrils. Posterior edge of the preorbital bone truncated obliquely—
Golden Grey Mullet *Liza aurata* (Fig. 57; Plate 46; p. 167)

— Scales on the dorsal side of the head extending to the nostrils or beyond. Posterior edge of the preorbital bone rounded or truncated vertically—
Thin-lipped Grey Mullet *Liza ramada* (Fig. 58; p. 167)

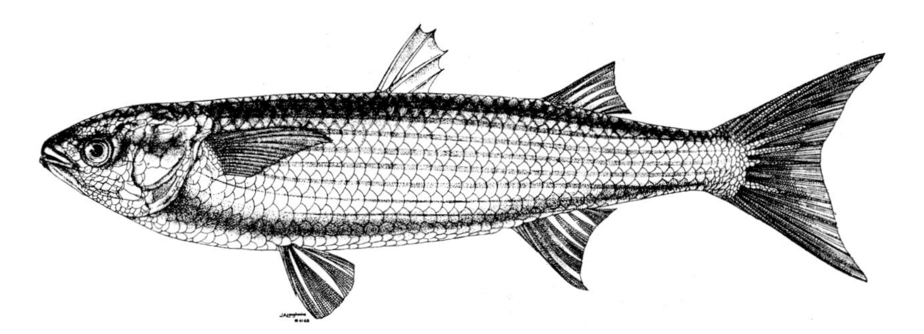

Fig. 57. Golden Grey Mullet *Liza aurata*.

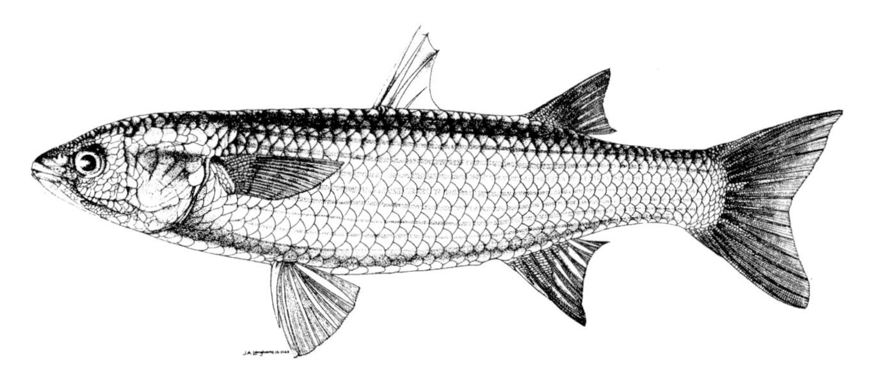

Fig. 58. Thin-lipped Grey Mullet *Liza ramada*.

Key to Stickleback Family GASTEROSTEIDAE

1 Less than 5 dorsal spines (normally 3), the longest about the same height as the dorsal fin. Gill openings narrow, restricted to the sides. Throat of the male vivid red during the breeding season—

Three-spined Stickleback *Gasterosteus aculeatus*
(Fig. 59; Plate 37; p. 168)

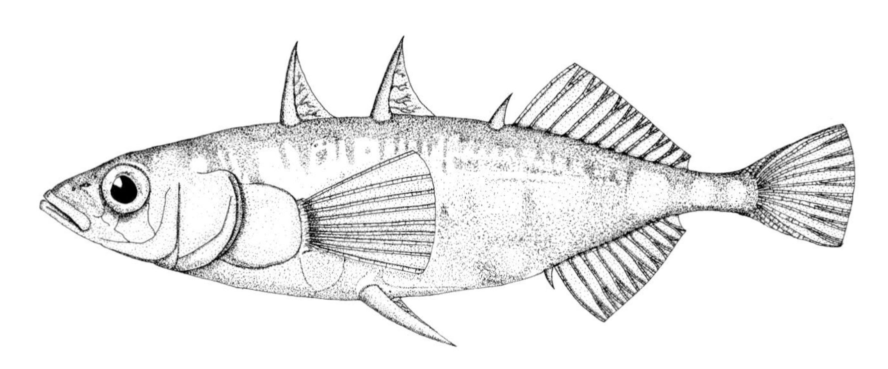

Fig. 59. Three-spined Stickleback *Gasterosteus aculeatus*.

— More than 6 dorsal spines (normally 9), the longest only about half the height of the dorsal fin. Gill openings wide, confluent ventrally. Throat of the male vivid black during the breeding season—

Nine-spined Stickleback *Pungitius pungitius*
(Fig. 60; Plate 38; p. 169)

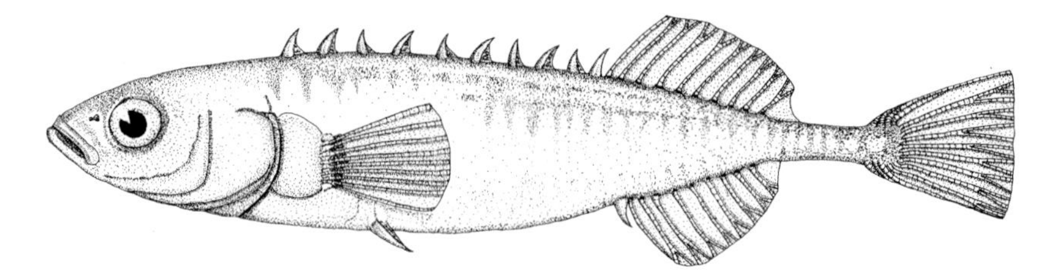

Fig. 60. Nine-spined Stickleback *Pungitius pungitius*.

Sculpin Family COTTIDAE

One species occurs in Britain (not found in Ireland)—
Common Bullhead *Cottus gobio* (Fig. 61; Plate 47; p. 170)

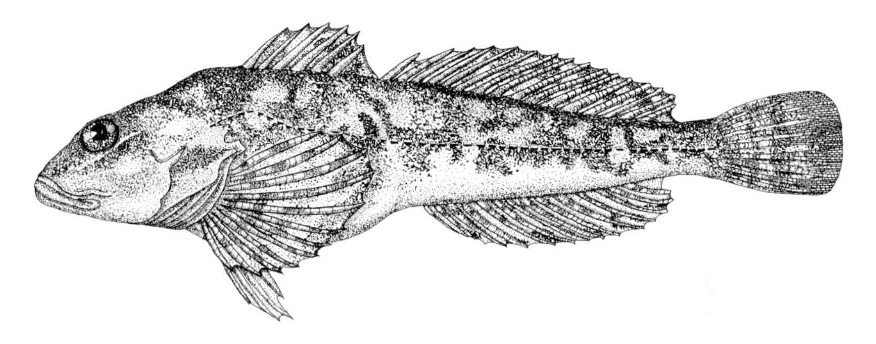

Fig. 61. Common Bullhead *Cottus gobio*.

Bass Family MORONIDAE

One species occurs in Britain and Ireland—
Sea Bass *Dicentrarchus labrax* (Fig. 62; Plate 39; p. 170)

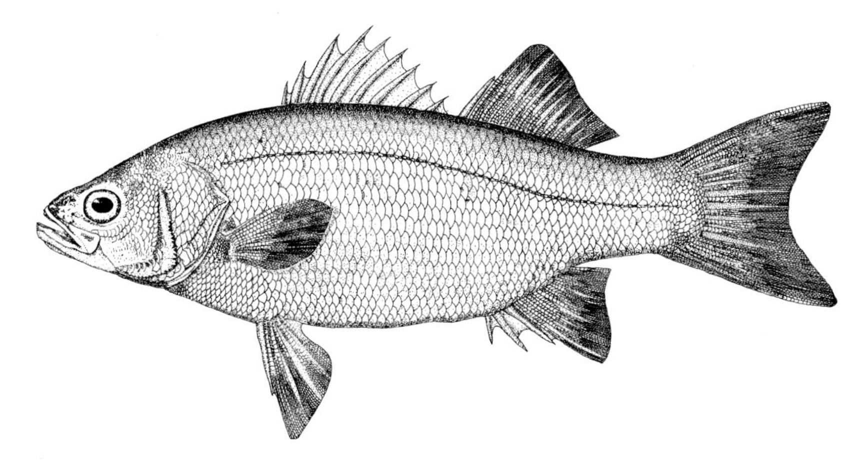

Fig. 62. Sea Bass *Dicentrarchus labrax*.

Key to Sunfish Family CENTRARCHIDAE

1 More than 50 scales along the lateral line. Dorsal fins almost separated by a notch. Length of the body more than three times its greatest depth—
Largemouth Bass ***Micropterus salmoides*** (Fig. 63; Plate 42; p. 172)

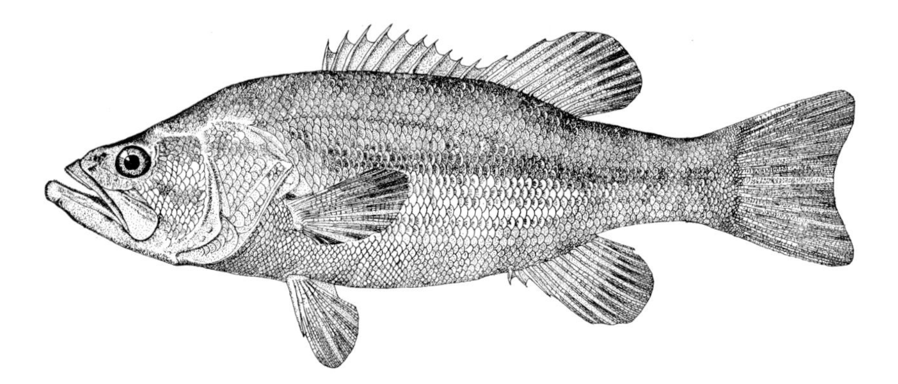

Fig. 63. Largemouth Bass *Micropterus salmoides*.

— Less than 50 scales along the lateral line. Dorsal fins continuous. Length of the body less than three times its greatest depth— **2**

2 More than 4 spines in the anal fin. Operculum ending in two flat points, neither of which is black in colour. Gill rakers long, the longest one longer than the diameter of the eye—
Rock Bass ***Ambloplites rupestris*** (Fig. 64; Plate 40; p. 171)

— Less than 4 spines in the anal fin. Operculum ending in a convex flap which is black in colour. Gill rakers short, the longest one shorter than the diameter of the eye—
Pumpkinseed ***Lepomis gibbosus*** (Fig. 65; Plate 41; p. 172)

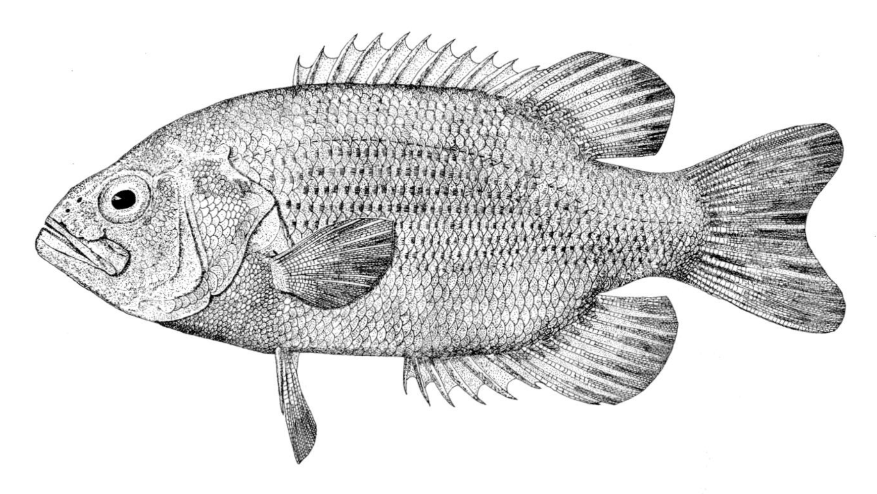

Fig. 64. Rock Bass *Ambloplites rupestris*.

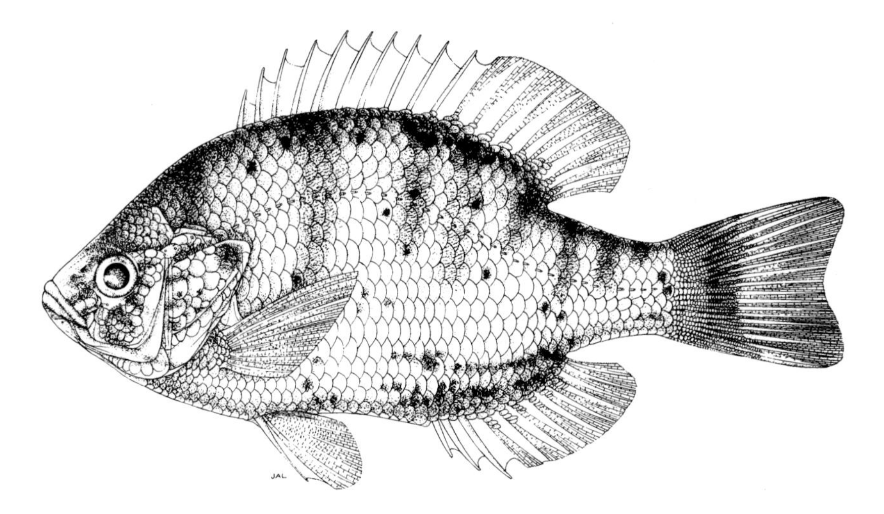

Fig. 65. Pumpkinseed *Lepomis gibbosus*.

Key to Perch Family PERCIDAE

1 Dorsal fins confluent. Less than 9 rays in the anal fin. Less than 50 scales along the lateral line—
Ruffe *Gymnocephalus cernuus* (Fig. 66; Plate 43; p. 173)

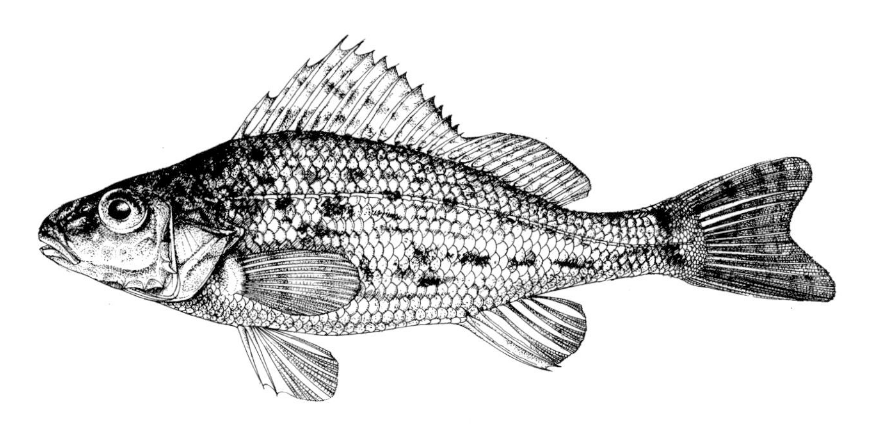

Fig. 66. Ruffe *Gymnocephalus cernuus*.

— Dorsal fins separate. More than 9 rays in the anal fin. More than 50 scales along the lateral line— **2**

2 Base of first dorsal fin longer than that of the second fin. Less than 70 scales along the lateral line. Short teeth only in the mouth—
European Perch *Perca fluviatilis*
(Fig. 67; Front cover illustration; p. 174)

— Base of first dorsal fin the same length as or shorter than that of the second fin. More than 70 scales along the lateral line. Both long and short teeth present in the mouth—
Pikeperch *Sander lucioperca* (Fig. 68; Plate 44; p. 175)

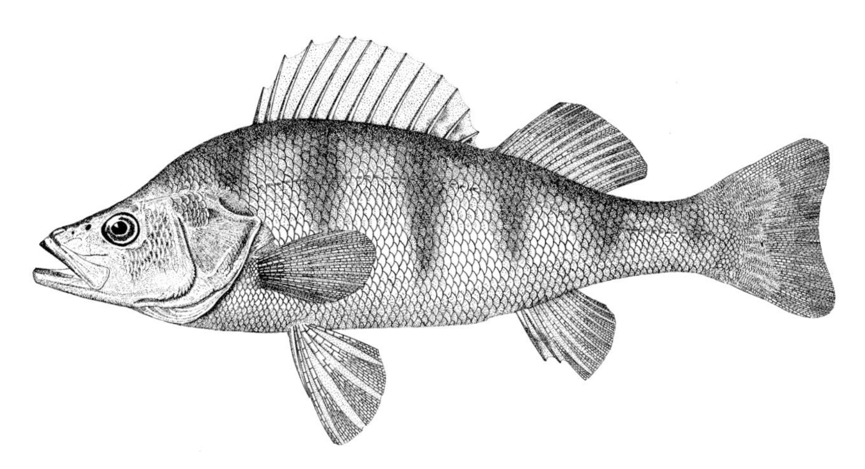

Fig. 67. European Perch *Perca fluviatilis*.

Fig. 68. Pikeperch *Sander lucioperca*.

Goby Family GOBIIDAE

One species occurs in Britain and Ireland—

Common Goby *Pomatoschistus microps* (Fig. 69; Plate 48; p. 175)

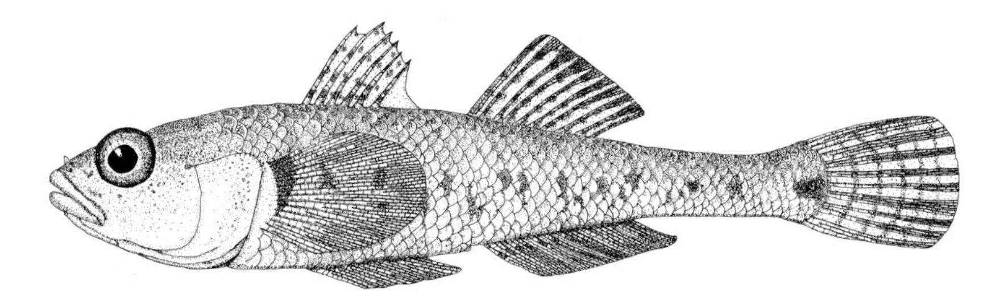

Fig. 69. Common Goby *Pomatoschistus microps*.

Flatfish Family PLEURONECTIDAE

One species occurs in Britain and Ireland—

Flounder *Platichthys flesus* (Fig. 70; p. 176).

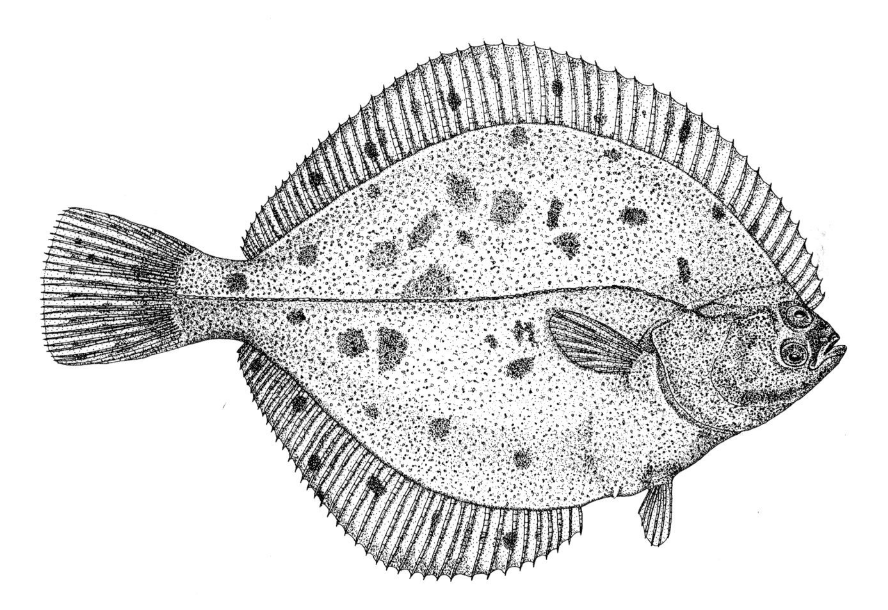

Fig. 70. Flounder *Platichthys flesus*.

COLOUR PLATES

Forty-eight species of fish are illustrated in colour Plates 1–48 on pages 89 to 112, including Channel Catfish (Plate 25) that has not been recorded from the wild but is sold in aquarium shops. The pictures are grouped to show fish that are mostly similar in general appearance, thereby providing a further aid to identification. European Perch is illustrated in colour on the front cover of this publication. European Eel and Smelt (Sparling) are illustrated in monochrome on pages 41 and 67 respectively.

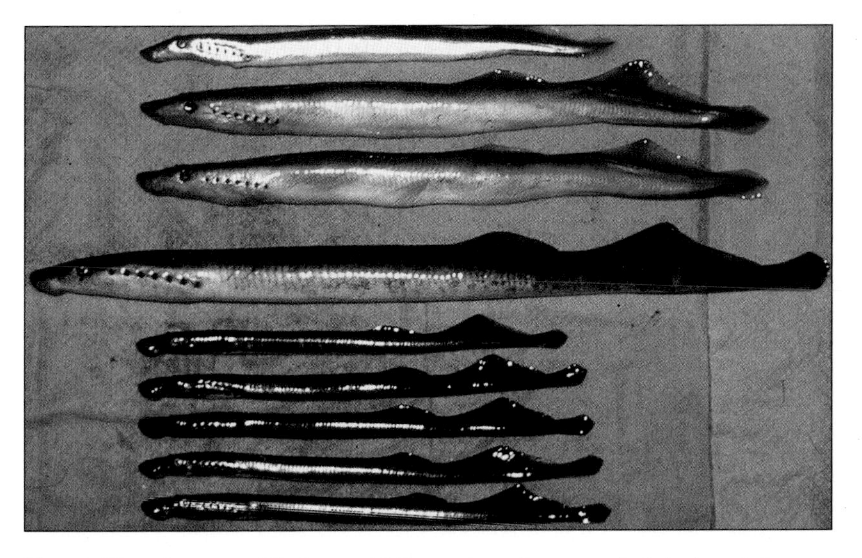

Plate 1. **River Lamprey.** The large middle specimen is a normal adult, those below are the dwarf form found only in Loch Lomond. The specimens above are the juvenile feeding phase, from the Forth Estuary (K. East).

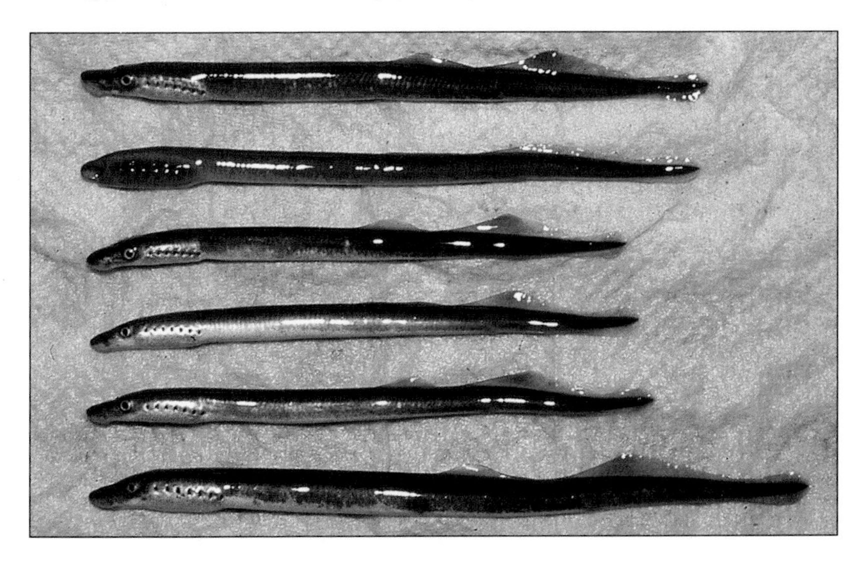

Plate 2. **Brook Lamprey.** All these lampreys are in the adult phase except for the specimen second from the top, which is a typical ammocoete larva with undeveloped sucker, eyes and tail fin (K. East).

Plate 3. **Sea Lamprey.** Typical large adult clearly showing the seven gill openings and the mottled back colouration (Environment Agency).

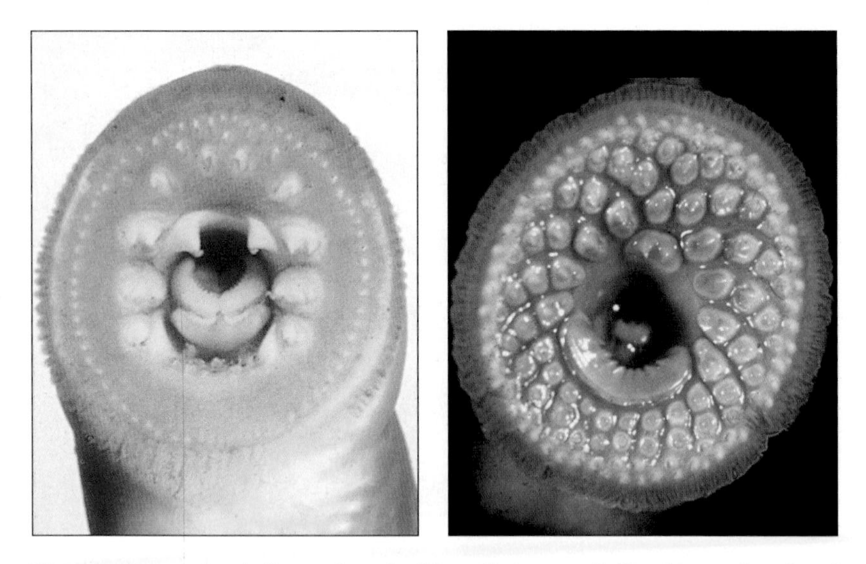

Plate 4. Lamprey oral disc and teeth: **River Lamprey** (left) with much reduced dentition compared to **Sea Lamprey** (right) (J. Clegg; P. S. Maitland).

Plate 5. **Allis Shad** (above) and **Twaite Shad** (below). Specimens from the Cree Estuary, caught in salmon nets (P. S. Maitland).

Plate 6. **Killarney Shad.** Three typical specimens of this rare subspecies of Twaite Shad, from Lough Leane (N. O. O'Maoileidigh).

Plate 7. **Silver Bream.** A fine specimen from the River Rhine (B. Hanfling).

Plate 8. **Crucian Carp.** A live specimen in an aquarium. Note the convex dorsal fin (Environment Agency).

Plate 9. **Roach.** From an original painting by David Lewins (© FBA).

Plate 10. **Rudd.** A typical adult specimen (Environment Agency).

Plate 11. **Common Minnow.** A mature adult male showing white breeding tubercles on its head (Environment Agency).

Plate 12. **Sunbleak.** A typical specimen of this recent alien introduction to Britain (T. Ostergaard).

Plate 13. **False Harlequin.** A live specimen of this recent alien introduction to Britain (M. Lorenzoni).

Plate 14. **Bitterling.** A live specimen in an aquarium, showing the characteristic lateral blue stripe (M. Hoult/C. Goldspink).

Plate 15. **Barbel.** A young specimen of this riverine species, showing typical barbels (Environment Agency).

Plate 16. **Common Gudgeon.** A typical adult, showing characteristic mottled colouring and mouth barbels (Environment Agency).

Plate 17. **Common Carp.** A fine angled specimen, showing characteristic scaling and barbels (R. Woods).

Plate 18. **Tench.** A young specimen in an aquarium (Environment Agency).

Plate 19. **Bleak.** A live specimen of this southern species in an aquarium (Environment Agency).

Plate 20. **Chub.** A typical specimen in an aquarium setting (Environment Agency).

Plate 21. **Orfe.** The golden form of Orfe, popular with aquarists and pondkeepers (Environment Agency).

Plate 22. **Dace.** A live specimen in an aquarium; the distribution of Dace is being extended through introductions by anglers (Environment Agency).

Plate 23. **Spined Loach.** A live specimen showing the typical lateral patterning and small barbels (M. Lorenzoni).

Plate 24. **Stone Loach.** A live specimen showing the diffuse mottled patterning (Environment Agency).

Plate 25. **Black Bullhead** (below) and **Channel Catfish** (above). Adult specimens caught in Lake Winnipeg, showing the characteristic barbels and body colouration (P. S. Maitland).

Plate 26. **Wels Catfish.** A fine specimen of this large species whose distribution is being extended by anglers throughout England (Environment Agency).

Plate 27. **Pollan.** Typical specimens of this endangered fish, from Lough Neagh (A. Ferguson).

Plate 28. **Grayling.** A fine specimen on a measuring board, clearly showing the large dorsal fin (Environment Agency).

Plate 29. **Vendace.** A typical specimen from Bassenthwaite Lake where this species is now threatened with extinction (P. S. Maitland).

Plate 30. **Powan.** A pair of fish from Loch Lomond – the female is plumper than the male (P. S. Maitland).

Plate 31. **Rainbow Trout.** The heavy black spotting is typical of this alien species. A characteristic pink lateral stripe along the body is not always well-developed and is not apparent in the particular strain illustrated here (Dept of Fisheries & Oceans, Canada).

Plate 32. **Brown Trout.** Typical wild specimens from a highland loch, showing the patterning of dark spots which is unique to each individual (P. S. Maitland).

Plate 33. **Arctic Charr.** Two brightly coloured male fish from Loch an t-Seilich. Note the white edging to the lower fins and the light spots on a darker background – the opposite to Brown Trout (P. S. Maitland).

Plate 34. **Brook Charr.** Four fine specimens of this North American species from a Scottish loch, showing the characteristic patterning (R. N. Campbell).

Plate 35. **Burbot.** Extinct now in Britain, this specimen was caught in Lake Winnipeg and shows the characteristic dark mottling on a light brownish background (P. S. Maitland).

Plate 36. **Pike.** A young fish in an aquarium, showing the characteristic streamlined shape and barred camouflage colouration (FBA).

Plate 37. **Three-spined Stickleback.** An adult fish from Lindean Reservoir, showing the characteristic dorsal spines (A. Buckham).

Plate 38. **Nine-spined Stickleback.** A male in full spawning colours with characteristic black throat and belly but contrasting white ventral spine (Environment Agency).

Plate 39. **Sea Bass.** A young specimen showing the two dorsal fins; the anterior fin is lowered and the posterior fin is raised (Environment Agency).

Plate 40. **Rock Bass.** A young specimen of this North American fish, showing the large mouth and characteristic dorsal fin (Dept of Fisheries & Oceans, Canada).

Plate 41. **Pumpkinseed.** An aquarium-bred specimen of this introduced species from North America, showing the erect finnage and characteristic black spot on the gill cover (P. S. Maitland).

Plate 42. **Largemouth Bass.** A young specimen of this North American species, showing the typical dorsal fin and body colouration (Dept of Fisheries & Oceans, Canada).

Plate 43. **Ruffe.** A specimen from Loch Lomond (where it was introduced by anglers about 1980), showing the characteristic mottled colouration on the body and fins (P. S. Maitland).

Plate 44. **Pikeperch.** A young fish in an aquarium, showing the typical fin and body patterning (Environment Agency).

Plate 45. **Thick-lipped Grey Mullet.** A typical specimen of this common fish, from the Cree Estuary (P. S. Maitland).

Plate 46. **Golden Grey Mullet.** A specimen netted from a brackish lake at Malltraeth, Anglesey (P. S. Maitland).

Plate 47. **Common Bullhead.** A typical adult specimen of this aggressive stream fish (Environment Agency).

Plate 48. **Common Goby.** An adult male in full spawning colours (P. J. Miller).

KEY TO EGGS OF FAMILIES

The eggs of various fish are often found during work in freshwater habitats, and it may be useful to know to which species they belong. The following key has been constructed as an aid in such identifications. It should be noted that it identifies eggs to family level only.

Ideally, eggs should be examined when fresh, for their colour may be an important character in identification. Furthermore, it is useful to know exactly where and when they were collected, as both the site of laying and the time of year are relevant and used in the key below. The attachment of the eggs to each other and to weeds, stones or other substrata are also relevant points to note. Further information is given later under **Reproduction** for each species in Notes on Distribution and Ecology on pages 133–176.

Although this key allows identification to family level only, it is often possible to conclude with reasonable certainty the exact species involved. Thus, for 14 of the families, only one species occurs in fresh water in the British Isles. In waters where all the species present are known, and there is only one species in the family concerned, then it is reasonable to assume specific identity of the eggs. In Ireland, the key given by Bracken & Kennedy (1967) can be used for some families. Certain species have never been known to breed in Britain or Ireland (e.g. *Acipenser sturio* and *Coregonus oxyrinchus*).

1 Eggs found only in the sea— ANGUILLIDAE, MUGILIDAE, GOBIIDAE, MORONIDAE, PLEURONECTIDAE

— Eggs found in fresh or brackish waters— **2**

2 Eggs single, or occasionally in small clumps of variable numbers (usually less than 10 eggs)— **3**

— Eggs in definite clumps or ribbons (usually more than 50 in each)— **18**

3 Eggs loose, slightly buoyant, usually in or near estuaries. (Diameter of eggs 4–5 mm. Occurring from May to July)— CLUPEIDAE

— Eggs not loose and buoyant— **4**

4 Eggs not adhesive, or only weakly so. Usually found (often buried) among coarse gravels in well oxygenated places; normally in running water or at the edges of clean lakes— **5**

— Eggs adhesive. Usually found attached to weed or stones (never buried) in a variety of habitats— **8**

5 Eggs less than 1.5 mm in diameter, white in colour. (Eggs buried among small stones and gravel in shallow depressions in running water. Occurring from March to July)— PETROMYZONTIDAE

— Eggs more than 1.5 mm in diameter, yellow-orange in colour— **6**

6 Eggs less than 3 mm in diameter. (Eggs found among coarse gravel and stones in lakes. Occurring from December to March)— COREGONIDAE

— Eggs more than 3 mm in diameter— **7**

7 Eggs less than 4 mm in diameter. (Eggs found buried among coarse gravel in running water. Occurring from March to May)— THYMALLIDAE

— Eggs more than 4 mm in diameter. (Eggs found buried in groups in coarse gravel in running water and at the edges of clean lakes. Occurring from October to March)— SALMONIDAE

8(4) Eggs more than 2 mm in diameter— **9**

— Eggs less than 2 mm in diameter— **14**

9 A nest or protection of some kind sheltering the eggs— **10**

— No nest nor protection sheltering the eggs— **12**

10 Eggs found within freshwater mussels (Unionidae). (Eggs yellow in colour. Occurring from May to July)— CYPRINIDAE (*Rhodeus sericeus* only)

— Eggs not occurring with freshwater mussels, but in a circular depression in the substratum from June to August— **11**

11 Eggs about 3 mm in diameter, with a gelatinous coat, pale cream in colour, laid in a mass in a circular depression excavated by the parents— ICTALURIDAE

— Eggs less than 3 mm in diameter, amber to pale yellow in colour, laid in a circular depression excavated by the parents— CENTRARCHIDAE

12(9) Eggs grey, more than 2.5 mm in diameter, attached to stones in rivers where the current is strong. (In Europe, eggs occur from April to July, but there are no records from rivers in Britain or Ireland)— ACIPENSERIDAE

— Eggs not grey, less than 2.5 mm in diameter, found attached to weeds in lakes and slow-flowing rivers— **13**

13 Eggs pale yellow, occurring from May to July— SILURIDAE

— Eggs brown, occurring from February to May— ESOCIDAE

14(8) Eggs found from January to February. (Diameter 0.8–1.5 mm, attached to plants and stones. No recent records from Britain)— GADIDAE

— Eggs found from March to September— **15**

15 Eggs less than 1.2 mm in diameter, rarely found in standing water— **16**

— Eggs more than 1.2 mm in diameter, found in both standing and running water— CYPRINIDAE (except *Rhodeus*)

16 Eggs found in or near estuaries, from March to April. Eggs yellow in colour (white when infertile), firmly attached to plants and stones— OSMERIDAE

— Eggs found in fresh water only, from May to July. Eggs whitish yellow in colour; attached to plants and stones— **17**

17 Eggs yellowish, less than 0.9 mm in diameter— COBITIDAE

— Eggs whitish, more than 0.9 mm in diameter— BALITORIDAE

18(2) Eggs white or whitish yellow, laid in ribbons among weed, or as a single large clump at the base of weeds— PERCIDAE

— Eggs yellow, laid in small clumps— **19**

19 Eggs more than 2 mm in diameter, laid in a single mass (usually more than 10 mm across) adhering to the undersurfaces of stones. (Eggs occurring from February to May, usually in streams)— COTTIDAE

— Eggs less than 2 mm in diameter, laid in a single mass (usually less than 10 mm across) inside a small nest built of pieces of vegetation. (Eggs occurring from March to July)— GASTEROSTEIDAE

KEY TO POST-LARVAE OF FAMILIES

The five main developmental stages found in freshwater fish in Britain and Ireland are as follows.

1. *Egg*. After fertilisation, this develops a secondary membrane.

2. *Larva*. The stage between hatching from the egg and final absorption of the yolk sac.

3. *Post-larva*. The stage between final absorption of the yolk sac and the completion of fin definition.

4. *Fry*. The stage between the completion of fin definition and the attainment of adult characteristics.

5. *Adult*.

Although it has been possible to define these stages more or less exactly, it should be noted that there is a gradual transition from fertilised egg to adult and thus a variety of intermediate stages are found. Balinsky (1948) defined more than 40 different stages between the fertilised egg and the fry stage in British Cyprinidae. The key given below has been developed for post-larvae, i.e. for a stage between fertilised egg and adult, keys for both of which have already been given in this volume. As with the keys to eggs and scales, this key to post-larvae identifies to family level only.

However, as with the other family keys, it is often possible to identify the exact species involved after the family has been ascertained. Thus with 14 of the families involved, only one species occurs in fresh water in Britain and Ireland. In addition, the distribution of some families is so restricted (e.g. Coregonidae) that only one species is likely to be present in any particular water. A key to all the species of Cyprinidae has been given by Balinsky (1948), whilst for Irish waters the key for coarse (non-salmonid) fish produced by Bracken & Kennedy (1967) can be used for some species. More recently, the key to the larval stages of non-salmonid fish (Pinder 2001) is a welcome addition to the FBA series of Scientific Publications and covers all individual species in the following families: Cyprinidae, Cobitidae, Balitoridae, Esocidae, Gasterosteidae, Cottidae and Percidae.

1 Found only in the sea— ANGUILLIDAE, MUGILIDAE, GOBIIDAE, MORONIDAE, PLEURONECTIDAE

— Found in fresh or brackish waters— **2**

2 No paired fins or lower jaw present. Seven pairs of gill openings. Usually found buried in sandy silt in running water— PETROMYZONTIDAE

— Both paired fins and lower jaw present. One pair of gill openings, each protected by an operculum— **3**

3 Distinct barbels present near the mouth— **4**

— Distinct barbels absent— **6**

4 Barbels anterior to the mouth. Obvious snout present— ACIPENSERIDAE

— Barbels posterior to the mouth. Obvious snout absent. Tadpole-like and pitch-black in colour. Often taken in large numbers due to batch guarding by parents— **5**

5 Four pairs of developing barbels— ICTALURIDAE

— Three pairs of developing barbels— SILURIDAE

6(3) Length of post-larva more than 12.5 mm— **7**

— Length of post-larva less than 12.5 mm— **10**

7 Dark stripe through the eye. Tail pointed at its end— ESOCIDAE

— No dark stripe through the eye. Tail rounded or concave at its end— **8**

8 Vent opening nearer the tail than the middle of the body. Found mainly in estuaries and the lower reaches of rivers— CLUPEIDAE

— Vent opening nearer the middle of the body than the tail— **9**

9 Length of post-larva more than 20 mm— SALMONIDAE

— Length of post-larva less than 20 mm— THYMALLIDAE

10(6) Vent opening nearer the anterior than the posterior end of the body— GADIDAE

— Vent opening midway or nearer the posterior than the anterior end of the body— **11**

11 Swim-bladder large, its length usually about twice that of the eye— **12**

— Swim-bladder small or absent, its maximum length never greater than that of the eye— **17**

12 Fine black marks on the edges of myotomes (muscle bands)— PERCIDAE

— No fine black marks on the edges of myotomes— **13**

13 Vent opening about midway along the body— CENTRARCHIDAE

— Vent opening in the posterior half of the body— **14**

14 Pigment cells scattered over most of the body— GASTEROSTEIDAE

— Pigment cells confined to specific parts of the body, often in linear series— **15**

15 Eyes small, less than half the length of the head— CYPRINIDAE

— Eyes large, more than half the length of the head— **16**

16 Length of post-larva more than 8 mm. Pigment cells in a double row down the back and round the gut— COREGONIDAE

— Length of post-larva less than 8 mm. Pigment cells not with the above pattern— OSMERIDAE

17(11) Length of post-larva more than 9 mm. Pectoral fins broader than long. Underside of the head smooth— COTTIDAE

— Length of post-larva less than 9 mm. Pectoral fins longer than broad. Underside of the head with numerous wart-like growths— **18**

18 Belly with a median line of pigment cells, those elsewhere starting to form patches in two lateral rows— COBITIDAE

— Belly without a median line of pigment cells, elsewhere pigment cells scattered and giving a mottled appearance— BALITORIDAE

KEY TO SCALES OF FAMILIES

As with their eggs, the scales of fish are often found during work in aquatic habitats and in some situations (e.g. studies of fish stomachs, the crops of birds or their nests, and lake sediments) they may be the only part of the fish left intact. It is clearly useful from an ecological viewpoint to be able to identify such scales, and the following key allows identification to family level for the scales of freshwater fish that occur in Britain and Ireland.

The key should be used with caution, and ideally several scales should be available for any particular identification. The characters referred to in the key are indicated below in Fig. 71.

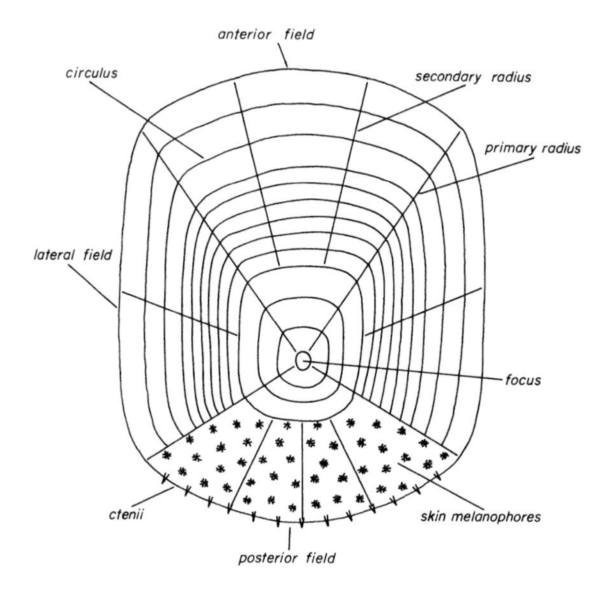

Fig. 71. General characteristics of fish scales. The (posterior) area exposed in life is indicated by pigment cells (melanophores) present in the overlying skin (the rest of the scale is normally covered by adjacent, overlapping, scales). The posterior field is either smooth-edged and rounded (cycloid scales) or rough and comb-like due to presence of ctenii (ctenoid scales, as shown above).

For examination, scales should be mounted on glass slides, either dry or in glycerine or some other mounting medium. The slides used can be normal microscope slides: if the scales are mounted dry then they should be sandwiched between two slides which are held firm by mounting tape. If mounted in a medium a normal cover slip should be placed on top. As well as microscope slides, two conventional projector slides can be used, with the scales mounted between them and the slides secured by tape. If mounted in this way, it is not necessary to have a microscope, for the slides can be projected and the images examined in the normal way.

Several freshwater species have no scales. In those which do possess them there are normally differences between scales from different parts of the body – those from the head region, the lateral line and adjacent to the fins showing differences in shape. The key below refers to typical body scales (always the great majority) from each of the families concerned, found on the main part of the body above and below the lateral line between the head and the tail. As with eggs, although the key identifies only to family level, it is often possible to conclude with reasonable certainty the exact species involved. Thus for 14 families only one species occurs in fresh water in the British Isles, and for some other families only one species may be present in the habitat under investigation.

Cycloid and ctenoid scales

The terms cycloid (rounded edge) and ctenoid (comb-like edge) specifically refer to the posterior edge of the scale – the part that is not covered by adjacent overlapping scales. The majority of families have cycloid scales which are smooth to the touch, but a few have ctenoid scales; these bear short, conical projections or ctenii (Fig. 71) and are rough to the touch. Ctenii are particularly numerous on the posterior field of Perch scales (see Fig. 78A). Bagenal (1978) mentions other differences between families with cycloid and ctenoid scales.

1 True scales absent or very much modified— PETROMYZONTIDAE, ACIPENSERIDAE, ICTALURIDAE, SILURIDAE, COTTIDAE, GASTEROSTEIDAE

— True scales covering most parts of the body (except in young fry)— **2**

2 Scales cycloid. Shape of entire scale ranging from rounded to subquadrate (Figs 72C–I) or longer than broad (Figs 72A,B)— **3**

— Scales ctenoid. Shape of entire scale either subquadrate (Figs 72J,K,M,N), oval with a rounded anterior edge (Fig. 72L), or longer than broad (Fig. 72B)— **13**

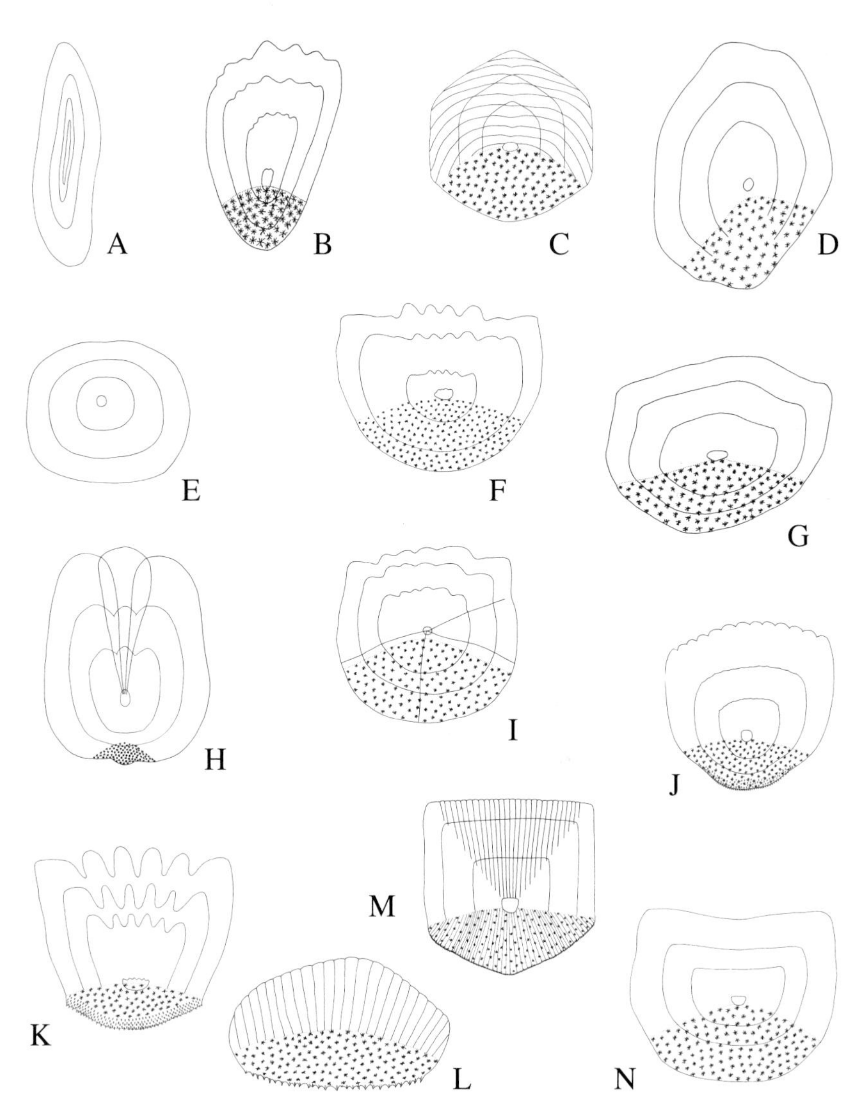

Fig. 72. Shapes of typical scales from members of the following fourteen families: A, Anguillidae; B, Pleuronectidae; C, Clupeidae; D, Salmonidae; E, Gadidae; F, Thymallidae; G, Coregonidae; H, Esocidae; I, Cyprinidae; J, Centrarchidae; K, Percidae; L, Gobiidae; M, Moronidae; N, Mugilidae. Note: Scales of Cobitidae, Balitoridae and Osmeridae are not shown.

3 Scales often asymmetrical and longer than broad— **4**

— Scales not as above— **5**

4 Circuli abnormal, modified to give a beaded appearance. Scale always greatly elongate— ANGUILLIDAE (Fig. 73A)

— Circuli normal. Anterior edge of scale scalloped in appearance— PLEURONECTIDAE* (Fig. 73B)

*Note: some scales on Flounder bear a few ctenii on the posterior field; these scales key out at couplet 13. Flounder scales are noticeably small in size.

5(3) Radii essentially transverse. Circuli also more or less transverse— CLUPEIDAE (Fig. 73C)

— Radii, when present, essentially radiating from the focus of the scale. Circuli never transverse— **6**

6 Primary radii absent— **7**

— Primary radii present— **11**

Fig. 73. *(On p. 125)*. Photographs and line-drawings of typical cycloid scales from: A, European Eel (Anguillidae); B, Flounder (Pleuronectidae); C, Twaite Shad (Clupeidae).

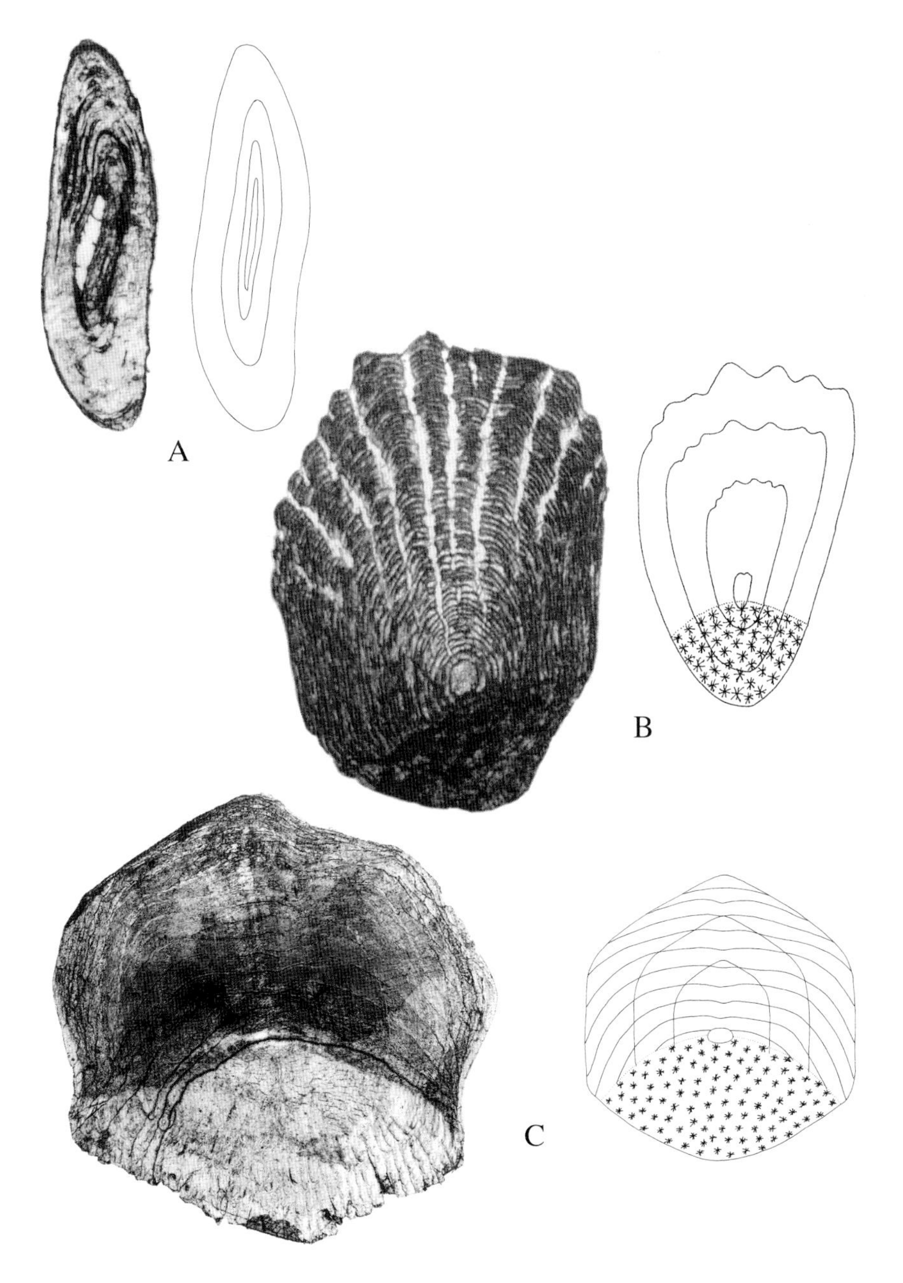
A
B
C

7 Scale circular. Circuli normally exactly concentric— GADIDAE (Fig. 74A)

— Scale never circular. Circuli never exactly concentric— **8**

8 Radii absent. Circuli not marking scale into four primary fields— **9**

— Secondary radii often present, usually in the posterior field. Circuli spaced and angulated to divide the scale into four primary fields— **10**

9 Focus of scale defined, more or less central, and usually longitudinally ovoid. Scale normally longer than wide— SALMONIDAE (Fig. 74B)

— Focus of scale poorly defined, displaced towards the anterior margin, and usually transversely ovoid. Scale normally wider than long— OSMERIDAE

10(8) Anterior edge and parts of circuli in the anterior field of the scale with three or more well marked ridges— THYMALLIDAE (Fig. 74C)

Fig. 74. *(On p. 127)*. Photographs and line-drawings of typical cycloid scales from: A, Burbot (Gadidae); B, Brown Trout (Salmonidae); C, Grayling (Thymallidae).

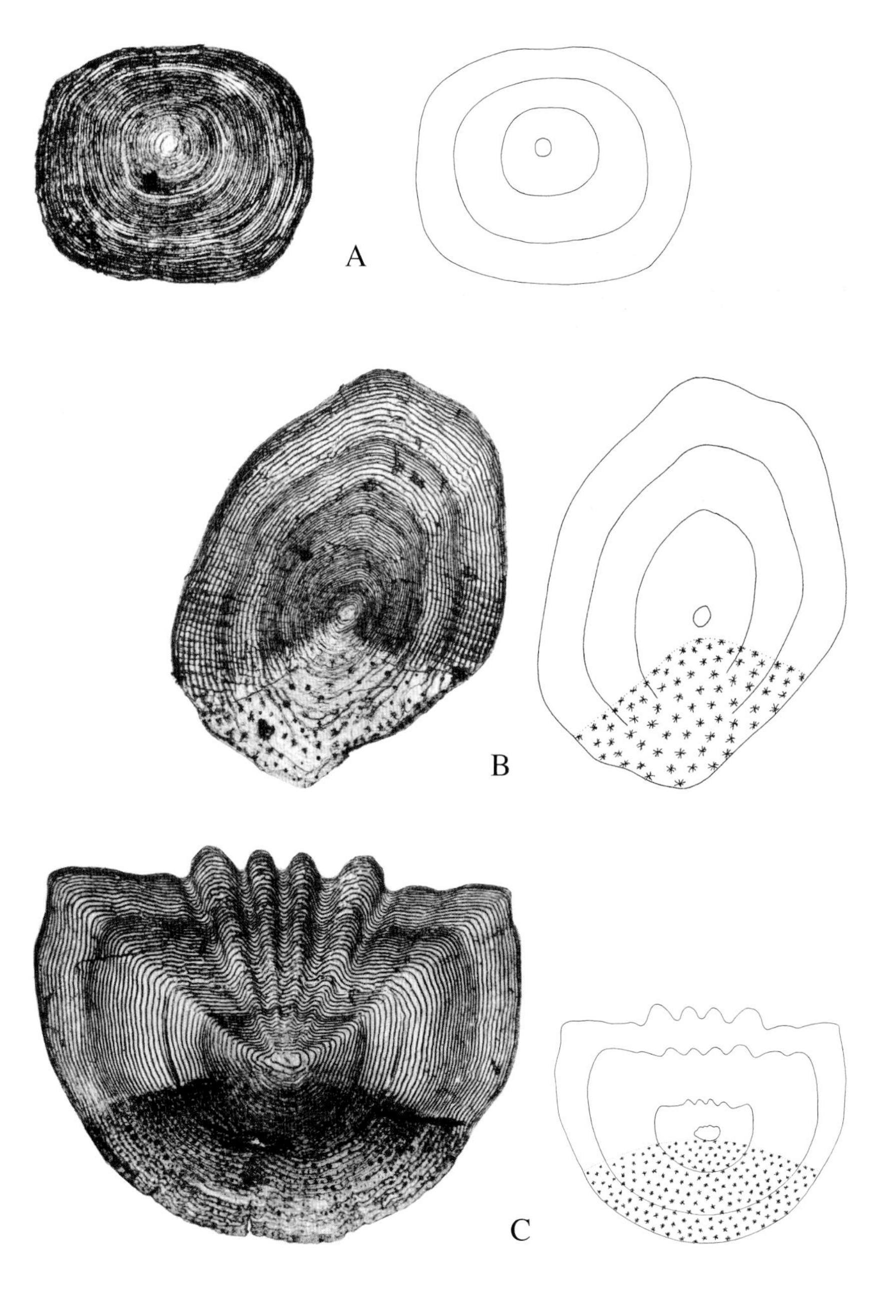
A
B
C

—(10) Anterior edge and parts of circuli in the anterior field of the scale without well marked ridges— COREGONIDAE (Fig. 75)

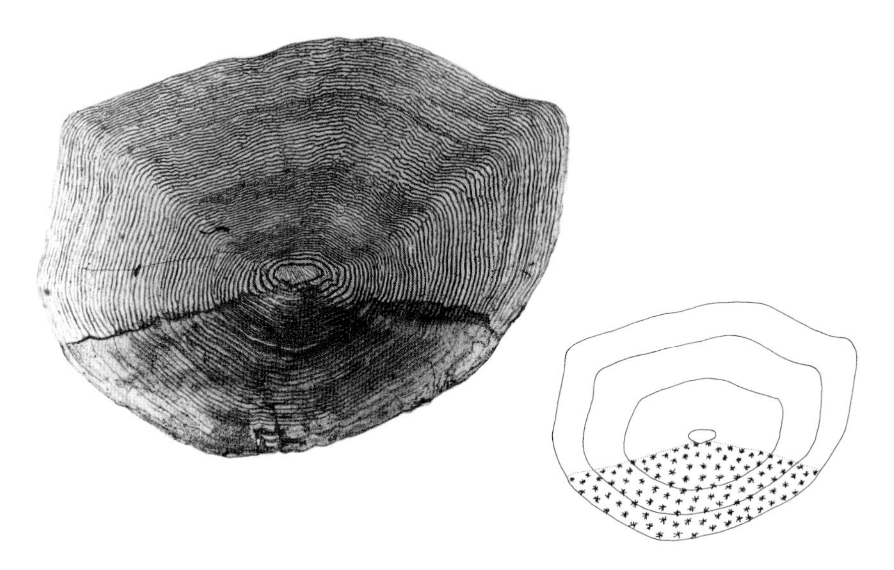

Fig. 75. Photograph and line-drawing of a typical cycloid scale from Powan (Coregonidae).

11(6) Primary radii found only in the anterior field— ESOCIDAE (Fig. 76A)

— Primary radii found in the posterior field and usually elsewhere also— **12**

12 Scale usually subquadrate in shape, never a rounded oval. Primary radii usually found in both anterior and posterior fields, but never in lateral fields. Scales usually large and approximately the same size as the eye of the fish— CYPRINIDAE (Fig. 76B)

— Scale a rounded oval in shape. Primary radii present in all fields. Scales always very small and much smaller than the eye of the fish— COBITIDAE and BALITORIDAE

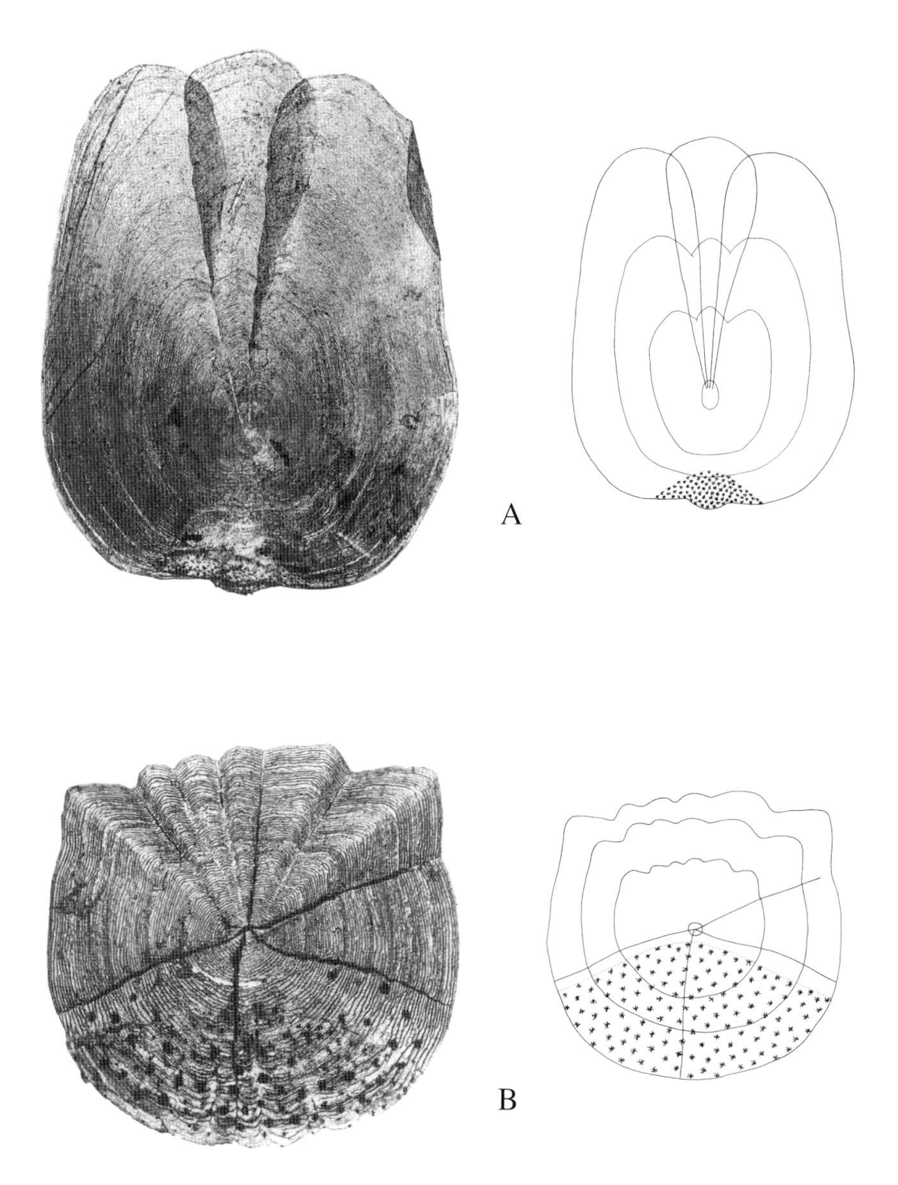

Fig. 76. Photographs and line-drawings of typical cycloid scales from: A, Pike (Esocidae); B, Crucian Carp (Cyprinidae).

13(2) Ctenii present in small numbers. Scale often asymmetrical and longer than broad— PLEURONECTIDAE
(cf. Figs 72B and 73B where ctenii are *absent*)

— Ctenii relatively numerous. Scale more or less symmetrical, rarely longer than broad— **14**

14 Ctenii occurring over much of the posterior field— CENTRARCHIDAE (Fig. 77)

— Ctenii occurring only in the marginal row, persisting elsewhere in the posterior field as basal segments of previous ctenii— **15**

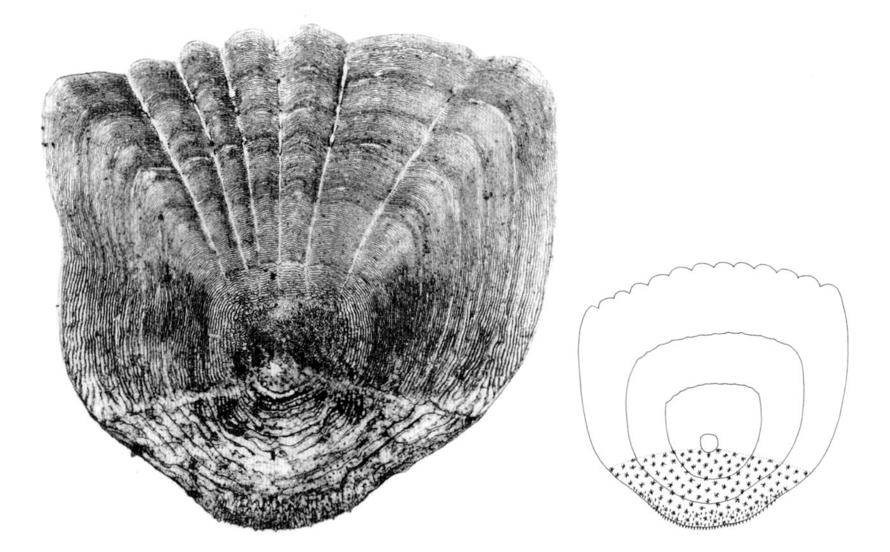

Fig. 77. Photograph and line-drawing of typical ctenoid scales from Largemouth Bass (left) and Rock Bass (right) (Centrarchidae).

15 Anterior edge of the scale conspicuously lobed— PERCIDAE (Fig. 78A)

— Anterior edge of the scale not conspicuously lobed— **16**

16 Scale much wider than long. Front margin of the scale rounded. Ctenii relatively long, usually more than 5% of length of the scale— GOBIIDAE (Fig. 78B)

— Scale about as wide as long. Front margin of the scale with well defined corners. Ctenii relatively short, usually less than 2% of length of the scale— **17**

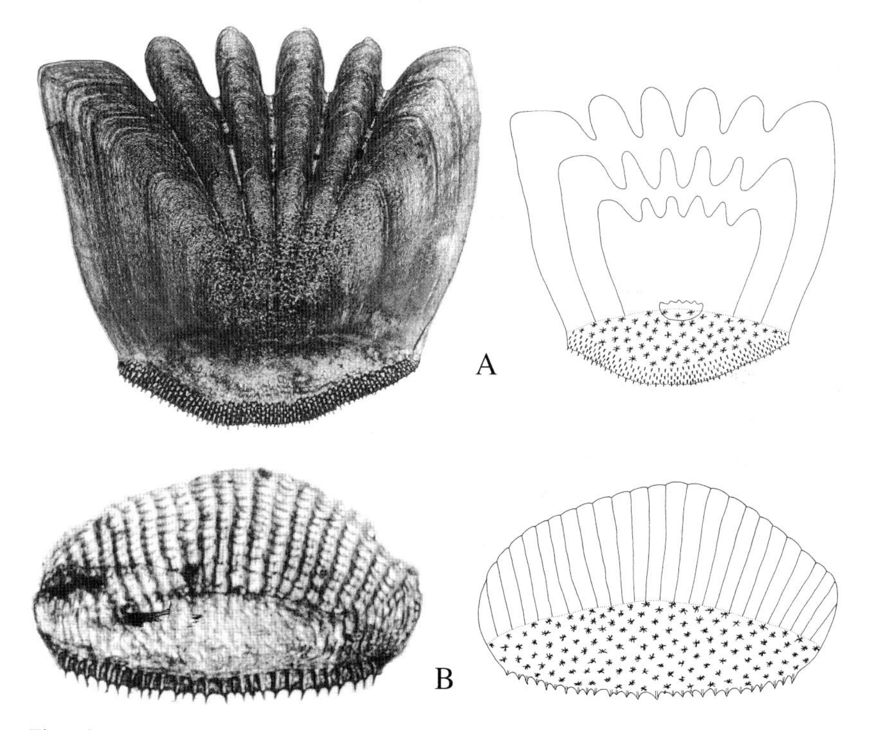

Fig. 78. Photographs and line-drawings of typical ctenoid scales from: A, European Perch (Percidae); B, Common Goby (Gobiidae).

17 Anterior corners of the scale acute. Anterior field with many primary and secondary radii— MORONIDAE (Fig. 79A)

— Anterior corners of the scale obtuse. Anterior field with few primary and secondary radii— MUGILIDAE (Fig. 79B)

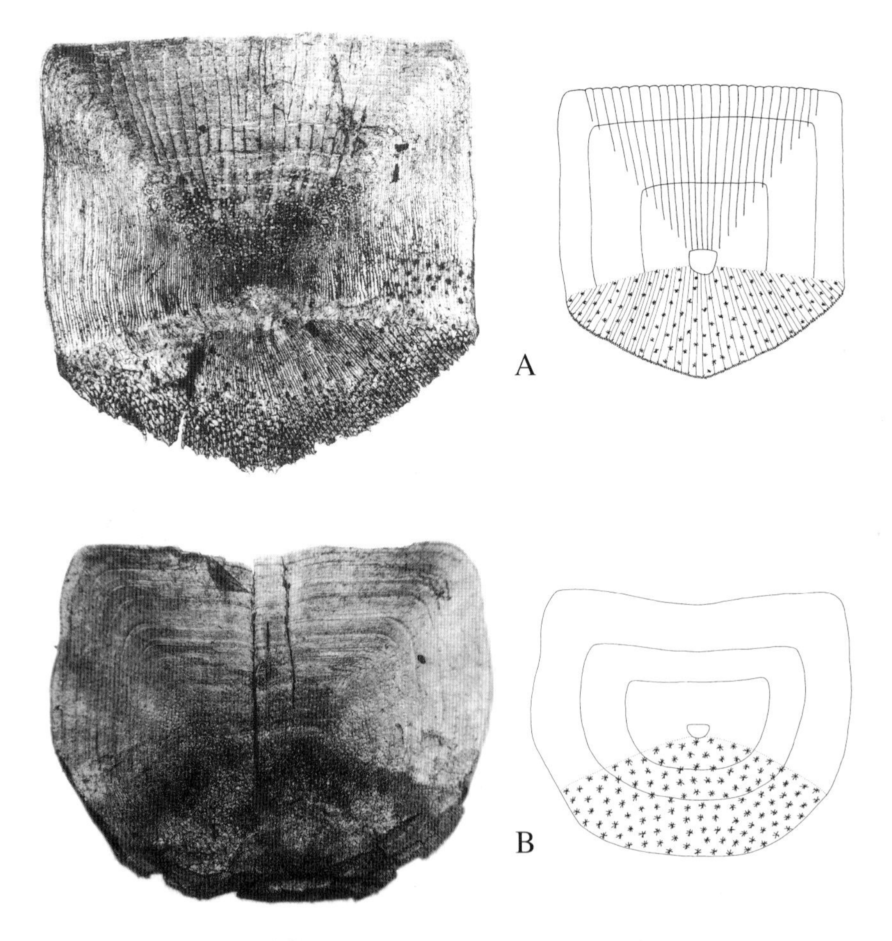

Fig. 79. Photographs and line-drawings of typical ctenoid scales from: A, Sea Bass (Moronidae); B, Thick-lipped Grey Mullet (Mugilidae).

NOTES ON THE DISTRIBUTION AND ECOLOGY OF EACH SPECIES
INCLUDING
DISTINCTIVE EXTERNAL FEATURES, CONSERVATION, AND LITERATURE

The following notes summarise important features of external morphology that are used for identification of species, and briefly outline the distribution and ecology of all the native and established alien species included in the preceding keys. In the ecological notes only the basic details concerning each species are given, in an abbreviated form. The information refers to the average conditions or measurements within Britain and Ireland, and not their extremes. Selected publications are indicated for each species and listed in full in the References on pages 209 to 239.

LAMPREY FAMILY: PETROMYZONTIDAE

River Lamprey *Lampetra fluviatilis* (Figs 6, 7, 9; Plates 1, 4)

Distinctive features: A wide maxillary plate is present, but no lower labial teeth; 7–10 teeth on the mandibular plate. Easily distinguished from other fish by the row of gill openings and, as adults, from other lampreys on body length. No scales. **Colour:** The ammocoete larvae are dull grey-brown, lighter underneath. The feeding phase is very silvery all over, but this darkens to a uniform dark olive-grey as the adults migrate upstream for spawning. A freshwater race in Loch Lomond (Scotland) is almost black at this stage. **Size:** 30–35 cm; maximum 50 cm and 150 g; the mean weight of females 32–34 cm long is about 62 g. The dwarf *praecox* form and the Loch Lomond race are smaller, some 18–24 cm. **Age:** Average adult age 5–7 years. **Distribution:** Occurs over much of western Europe from southern Norway to the western Mediterranean, in the sea and in accessible rivers. Indigenous to the British Isles, where it is found in most accessible river systems south of the Great Glen in Scotland (Map p. 180). There are several purely freshwater populations in Europe, but apparently only one in Britain – in the catchment of Loch Lomond. **Habitat:** Estuaries and accessible lakes, rivers and streams where there are adequate stretches of gravel and small stones in flowing water for spawning and beds of sandy silt in quieter water for the larvae. **Reproduction:** April–May. Clear eggs (1 mm) laid in nests built by the adults among stones in running water. The ammocoete larvae live in silted streambeds for 3–5 years, then metamorphose at about 10–12 cm into young adults which migrate to the sea (or a large lake). After 12–18 months, the mature adults migrate upstream again in winter and spring. 19,000–20,000 eggs per female. All adults die after spawning. **Food:** Filtered organic material when larvae, fish when adult. **Value:** An important commercial species in Scandinavia, the adults being caught in nets and traps as they migrate upstream. Many are smoked or grilled and eaten whole, for the gut is empty during migration and there is no bone in these cartilaginous fish. Of no angling value but there is a commercial fishery on the Yorkshire Ouse which takes about two tonnes each year for use as bait by anglers. **Conservation:** Because of its decline, the River Lamprey is now given some protection. It is listed in Annexes IIa and Va of the EU Habitats and Species Directive and in Appendix III of the Bern Convention, and as a Long List Species in the UK Biodiversity Action Plan. There is no Red Data Book for fish in Britain, nor is this species listed in the 1981 Wildlife and Countryside Act, but Maitland (2000) considers this species to be Vulnerable. The Red Data Book for Ireland (Whilde (1993), published before the IUCN (1994) revision of categories, lists the River Lamprey as Indeterminate. **Literature:** Abou-Seedo & Potter (1979), Baxter (1954), Bird & Potter (1979a,b), Claridge et al. (1973), Hagelin (1959), Hardisty (1961a,b, 1964), Hardisty & Huggins (1970), Hardisty & Potter (1971), Hardisty et al. (1970), Huggins & Thompson (1970), Kainua & Valtonen (1980), Maitland (1980a,b), Maitland et al. (1984, 1994), Morris (1989), Ojutkangas et al. (1995), Sjoberg (1980), Tuunainen et al. (1980), Valtonen (1980).

Brook Lamprey *Lampetra planeri* (Figs 6, 7, 10; Plate 2)

Distinctive features: A wide maxillary plate is present but no lower labial teeth; 5–9 blunt teeth on the mandibular plate. Easily distinguished from other fish by the row of gill openings and, as adults, from other lampreys on body length. No scales. **Colour:** The ammocoete larvae are mainly dull grey-brown, though lighter underneath. A golden form does occur occasionally. **Size:** 10–15 cm; maximum 18 cm. Fully grown ammocoetes are larger than the adults for there is weight loss between metamorphosis and spawning. **Age:** Average adult age 6–7 years. **Distribution:** A purely freshwater species occurring over much of western Europe, especially in basins associated with the Baltic and North Seas. Indigenous to the British Isles, where it is fairly common except in the north and west of Scotland and most islands (Map p. 180). **Habitat:** Clear rivers and streams where there are adequate stretches of gravel and small stones in flowing water for spawning and sandy silt beds in quieter water for the larvae. **Reproduction:** April–June. Clear eggs (0.9 mm) laid in nests in running water; these are depressions created by the adults removing stones with the help of their suckers. Larvae live in burrows in sandy silt in streams and rivers, metamorphosing after 3–5 years to adults which migrate upstream, spawn and die soon afterwards. 850–1,400 eggs per female. **Food:** Filtered organic material (e.g. diatoms and other algae and bacteria-coated detritus) when larvae. The adults do not feed. **Value:** Of no commercial value, but the larvae are occasionally used as bait for angling. **Conservation:** Because of its decline, the Brook Lamprey is now given some protection. It is listed in Annex IIa of the EU Habitats and Species Directive and in Appendix III of the Bern Convention, and as a Long List Species in the UK Biodiversity Action Plan. There is no Red Data Book for fish in Britain, nor is this species listed in the 1981 Wildlife and Countryside Act, but Maitland (2000) considers this species to be Vulnerable. The Red Data Book for Ireland (Whilde (1993), published before the IUCN (1994) revision of categories, lists the Brook Lamprey as Indeterminate. **Literature:** Abakumov (1964), Baxter (1954), Bird & Potter (1979a,b), Eneqvist (1938), Hardisty (1944, 1961a,b,c, 1964, 1970), Hardisty & Potter (1971), Hardisty et al. (1970), Huggins & Thompson (1970), Lohnisky (1966), Maitland (1980a), Maitland et al. (1994), Malmqvist (1978, 1980a,b), Moore & Potter (1976), Potter & Osborne (1975).

Sea Lamprey *Petromyzon marinus* (Figs. 7, 8; Plates 3 and 4)

Distinctive features: A characteristic tooth pattern; adults with very mottled colour patterning; the largest European lamprey. Easily distinguished from other fish by the row of gill openings and, as adults, from other lampreys on body length. No scales. **Colour:** The ammocoete larvae are dark greyish brown above and light grey below, whereas the adults are brownish grey with extensive black mottling. Some animals become golden brown, almost orange, at spawning time. **Size:** 50–70 cm; maximum 86 cm; a 53 cm specimen weighs about 360 g. **Age:** Average adult age 7–9 years. **Distribution:** A native anadromous species common along most of the Atlantic coastal area of western Europe and eastern North America. Indigenous to the British Isles, where it is found mostly in large rivers south of the Great Glen in Scotland (Map p. 181). **Habitat:** The sea, estuaries and accessible rivers. The riverine habitat must include adequate stretches of gravel and small stones in flowing water for spawning and sandy silt beds in quieter water for the larvae. **Reproduction:** June–July. Clear eggs (1.2 mm) are laid in nests created as depressions by moving stones and gravel in running water; these hatch into elongated blind ammocoete larvae which live buried in sandy silt in quiet waters for 2–5 years. They then metamorphose and migrate to the sea where they mature, returning to fresh water to spawn after 3–4 years. 34,000–240,000 eggs per female. **Food:** Filtered organic material (diatoms, bacteria, etc.) when larvae, fish (including Basking Sharks and other large fish) when adult. **Value:** Formerly more popular than now, it is still trapped or netted commercially in some European rivers when ascending to spawn (e.g. in Portugal). Never angled for. A serious pest to commercial and sport fisheries in the Great Lakes of North America. **Conservation:** Because of its decline, the Sea Lamprey is now given some protection. It is listed in Annex IIa of the EU Habitats and Species Directive and in Appendix III of the Bern Convention, and as a Long List Species in the UK Biodiversity Action Plan. There is no Red Data Book for fish in Britain, nor is this species listed in the 1981 Wildlife and Countryside Act, but Maitland (2000) considers this species to be Vulnerable. The Red Data Book for Ireland (Whilde (1993), published before the IUCN (1994) revision of categories, lists the Sea Lamprey as Indeterminate. **Literature:** Applegate (1950), Beamish & Medland (1988), Bird et al. (1994), Hardisty (1969), Hardisty & Potter (1971), Lelek (1973), Maitland (1980a), Manion (1967), Manion & McLain (1971), Morman et al. (1980), Newth (1930).

STURGEON FAMILY: ACIPENSERIDAE

Common Sturgeon *Acipenser sturio* (Fig. 11)

Distinctive features: Snout pointed; no scales but lateral scutes (24–39) less than twice as high as broad. Also known as Atlantic Sturgeon. It is important to check the identification of any specimens taken in Britain or Ireland for, although *Acipenser sturio* was originally the only species found here, escapes of both *Acipenser baeri* and *Acipenser ruthenus* have been found in recent years. **Colour:** Usually dark bluish brown on the back, lightening down the sides to pale whitish yellow ventrally. **Size:** 1.5–2.5 m; maximum 3.45 m –

this fish weighed 320 kg. **Age:** Average adult age 20–50 years. **Distribution:** Formerly much of western Europe, including the Mediterranean and Black Seas. Indigenous to the British Isles, but although it has been found in the lower reaches of several large rivers it has never been recorded as spawning here. Now extremely rare. **Habitat:** The open sea, coastal waters, estuaries and large accessible rivers. **Reproduction:** April–July in the middle reaches of a few European rivers (e.g. River Dordogne). The grey eggs (2.3 mm) adhere to stones in running water and hatch after 3–5 days. The young move to the sea after 2 years, except for purely freshwater populations (e.g. in Lake Ladoga in Russia). Adults mature after 7–15 years and may live for up to 50 years. 800,000–2,400,000 eggs per female. **Food:** Benthic invertebrates (collected by the use of sensitive barbels and a protrusible sucking mouth) and, when adult, some fish. **Value:** Formerly very important commercially and caught in large numbers across Europe for flesh and caviar. Traditionally, fish caught around Britain were offered to the Monarch. This species is now almost extinct and is one of Europe's most threatened fish. **Conservation:** Because of its severe decline, the Common Sturgeon is now given some protection. It is listed in Schedule 5 of the 1981 Wildlife and Countryside Act, in Annexes IIa and Va of the EU Habitats and Species Directive, in Appendix II of the Bern Convention, and as a Priority Species in the UK Biodiversity Action Plan. There is no Red Data Book for fish in Britain, but Maitland (2000) considers this species to be Critically Endangered. The Red Data Book for Ireland (Whilde 1993) does not deal with this species. **Literature:** Almaca (1988), Barannikova (1987), Castelnaud et al. (1991), CEMAGREF (1997), Lepage & Rochard (1995), Rochard et al. (1990, 1997), Smith (1985), Van Eenennaam & Doroshove (1998), Williot (1991), Williot et al. (1997).

EEL FAMILY: ANGUILLIDAE

European Eel *Anguilla anguilla* (Fig. 12)

Distinctive features: Elongate, cylindrical body with small gill openings; one pair of pectoral fins, but no pelvic fins; minute cycloid scales embedded in the skin. **Colour:** For most of their lives in fresh water, eels are dark brown above and yellowish below (Yellow Eels), but on maturation during their downstream migration they assume a silvery grey colouring (Silver Eels). **Size:** 40–90 cm; maximum about 2 m; British rod record 5.046 kg (1978:

Kingfisher Lake). **Age:** Average adult age 10–20 years. **Distribution:** Western Europe, including the Mediterranean and Black Seas. Indigenous to the British Isles, where it is widespread and common in most catchments (Map p. 181). **Habitat:** A wide variety of freshwater habitats accessible from the sea – ditches, streams, rivers, canals, ponds, lakes and reservoirs. **Reproduction:** Adults spawn in the Sargasso Sea from March to June, laying clear eggs (1.0 mm). Larvae (leptocephali) drift across the Atlantic in the Gulf Stream for one or more years before reaching Europe; they then migrate into fresh water as elvers, gradually maturing as Yellow Eels at 8–18 years. These then migrate, as Silver Eels, back to the Sargasso Sea. Each female has several million eggs. **Food:** Invertebrates (especially molluscs and crustaceans) and fish. **Value:** Of commercial importance in many countries, including Britain and Ireland, where it is caught in traps, nets and by baited hooks. Often smoked. Of sporting value in a number of areas. Numbers have declined alarmingly in some areas in recent years and control over commercial exploitation is becoming increasingly necessary. **Conservation:** Though apparently undergoing decline across much of Europe, this native species is still relatively common and there are no national conservation measures at present. **Literature:** Bertin (1956), Carss et al. (1999), Colombo et al. (1984), Costa et al. (1992), Desaunau & Guerault (1997), Frost (1945), Kennedy (1984b), McLeave (1980), Moore & Moore (1976) Moriarty (1973, 1978, 1983), Naismith & Knights (1990, 1993), Pilcher & Moore (1993), Sadler (1979), Schmidt (1922), Sinha & Jones (1975), Tucker (1959).

HERRING FAMILY: CLUPEIDAE

Allis Shad *Alosa alosa* (Figs 13, 14; Plate 5)

Distinctive features: A well developed median notch in the upper jaw; 85–130 long gill rakers; no vomerine teeth; more than 70 lateral cycloid scales. **Colour:** Very silvery, with the back greenish blue, lightening down the sides to white on the belly. A few large dark spots are sometimes present. **Size:** 30–50 cm; maximum 70 cm; British rod record 2.167 kg (1977: Chesil Beach, Dorset) – but it is no longer legal to fish for this species in the UK. **Age:** Average adult age 4–8 years. **Distribution:** Western Europe from southern Norway to Spain and in the Mediterranean eastward to northern Italy. Indigenous to the British Isles, where it was supposed to breed in some southern rivers, but although numbers are found in various coastal areas and occasional mature specimens in some larger rivers, it apparently no longer reproduces here (Map p. 182). **Habitat:** The sea, coastal waters and estuaries. Migrates well upstream to spawn in a few large accessible European rivers (e.g. the Rivers Loire and Garonne). **Reproduction:** May–June, the clear eggs (4.4 mm) are laid over spawning gravels in flowing water during fast chases by males following the females. The eggs hatch after 6–8 days and the young move downstream to the sea after 10–20 months. The adults mature after 3–4 years when they migrate, often in large numbers, back to the parent rivers. Hybridises with Twaite Shad. **Food:** Invertebrates, especially pelagic crustaceans, but also small fish. **Value:** Formerly more important commercially than now, it is still netted in some estuaries and the lower reaches of large rivers in continental Europe. **Conservation:** Because of recent decline, the Allis Shad is now given considerable legal protection. It is listed in Annexes IIa and Va of the EU Habitats and Species Directive, Appendix III of the Bern Convention, Schedule 5 of the 1981 Wildlife and Countryside Act, and as a Priority Species in the UK Biodiversity Action Plan. There is no Red Data Book for fish in Britain, but Maitland (2000) considers this species to be Critically Endangered. The Red Data Book for Ireland (Whilde 1993), published before the IUCN (1994) revision of categories, lists the Allis Shad as Endangered. **Literature:** Belaud et al. (1985, 1991), Boisneau et al. (1985, 1990, 1992), Collares-Pereira et al. (1999), De Groot (1990), Eiras (1981, 1983), Ellison (1935), Menneson-Boisneau et al. (1986), Prouzet et al. (1994), Taverny (1990), Taverny & Elie (1988), Wheeler et al. (1975).

Twaite Shad *Alosa fallax* (Figs 13, 15; Plates 5, 6)

Distinctive features: A marked median notch in the upper jaw; 30–80 short gill rakers; no vomerine teeth; less than 70 lateral cycloid scales. **Colour:** Very silvery in appearance with a greenish blue back, lightening down the sides to white on the belly. Several large dark spots are usually present. **Size:** 25–40 cm; maximum 55 cm; British rod record 1.247 kg (1978: Garlieston, Wigtownshire). **Age:** Average adult age 5–9 years. **Distribution:** Most of western Europe from southern Norway to the eastern Mediterranean. Indigenous to the British Isles and found along the coasts. Spawning populations occur in several rivers, especially around the Severn area but as far north as the Solway Firth (Map p. 182). **Habitat:** The sea, estuaries and large accessible rivers (e.g. Severn and Loire). Several purely freshwater populations occur in parts of Europe, some of which have been given subspecific status (e.g. *Alosa fallax killarnensis* in Loch Leane, near Killarney in Ireland; Plate 6). **Reproduction:** May–June, over gravel beds in the lower reaches and estuaries of rivers. The clear eggs (4.4 mm) hatch after 5–8 days and the young migrate downstream to the sea over the next few months, except in purely freshwater populations where the young remain in large lakes (e.g. Leane, Como, Lugano and Maggiore). The adults mature after 3–4 years. 75,000–200,000 eggs per female. Hybridises with Allis Shad. **Food:** Mainly invertebrates, especially pelagic crustaceans. **Value:** Though less common than formerly, it is still netted commercially in some parts of Europe. Though not the target species, considerable numbers are also caught by anglers in various rivers in England such as the Rivers Usk and Wye. **Conservation:** Because of recent decline, the Twaite Shad is now given some legal protection. It is listed in Annexes IIa and Va of the EU Habitats and Species Directive, Appendix III of the Bern Convention, and as a Priority Species in the UK Biodiversity Action Plan. There is no Red Data Book for fish in Britain, but Maitland (2000) considers this species to be Endangered. The Red Data Book for Ireland (Whilde 1993), published before the IUCN (1994) revision of categories, lists the Twaite Shad as Vulnerable. **Literature:** Aprahamian (1981, 1985, 1988, 1989), Assis et al. (1992), Claridge & Gardner (1978), Kennedy (1981), O'Maoileidigh et al. (1988), Trewavas (1938), Went (1953).

CARP FAMILY: CYPRINIDAE

Silver Bream *Abramis bjoerkna* (Figs 18, 34; Plate 7)

Distinctive features: Body deep and strongly compressed; 40–45 cycloid scales along the lateral line; dorsal fin with 8–9 and anal fin with 19–24 branched rays. Formerly known as *Blicca bjoerkna*. **Colour:** The back is dark olive-grey, grading into silvery white on the sides and into white on the belly. **Size:** 20–30 cm; maximum 35 cm; maximum weight 1.25 kg; British rod record 425 g (1988: Grime Spring). **Age:** Average adult age 5–10 years. **Distribution:** Found across much of central and northern Europe from eastern England to the Caspian Sea. Indigenous to southeast England but has been introduced to a few other catchments to the south and west (Map p. 183). Absent from Ireland. **Habitat:** Rich lakes, canals and slow-flowing rivers. **Reproduction:** May–July, communal spawning when the adhesive yellow eggs (1.9 mm) are laid among plants in shallow water. The young mature after

3–5 years and may live for up to 10 years. 11,000–82,000 eggs per female. Hybridises with Common Bream, Roach and Rudd. **Food:** Zooplankton when young; when older, invertebrates, especially worms, molluscs and insect larvae. **Value:** Of some local importance commercially (in net and trap fisheries) in parts of Europe and as a sport species in England and elsewhere. **Conservation:** This native species, though local, is relatively common and there are no national conservation measures at present. **Literature:** Hartley (1947a,b), Mooij (1989), Spataru (1968), Swinney & Coles (1982).

Common Bream *Abramis brama* (Figs 18, 35)

Distinctive features: Body very deep and compressed laterally; 51–60 cycloid scales along the lateral line; dorsal fin with 8–10 and anal fin with 24–30 branched rays. **Colour:** Dark brownish grey on the back, graduating to bronze-silver on the sides and yellowish silver on the belly. **Size:** 30–50 cm; maximum 80 cm; British rod record 7.541 kg (1999: Undisclosed Water). **Age:** Average adult age 6–12 years. **Distribution:** Found throughout much of Europe from Ireland to the Aral Sea. Indigenous to England, where it is fairly widespread and common, but introduced to Ireland, Wales and Scotland (Map p. 183). **Habitat:** Slow-flowing rivers and rich lakes. It also occurs in estuarine and brackish waters in some parts of Europe (e.g. Gulf of Finland). **Reproduction:** May–July, communal spawning at night when the yellow eggs (1.5 mm) are laid among weeds in shallow water. The adhesive eggs hatch in 5–10 days and the young mature after 3–5 years. 104,000–587,000 eggs per female. Hybridises with Silver Bream, Orfe, Roach and Rudd. **Food:** Uses an extendible tubular mouth to feed on benthic invertebrates, especially worms, molluscs and insect larvae. **Value:** Of considerable commercial value in central Europe where large numbers are caught in nets and traps. It is also important as a sport species in many countries, including Britain and Ireland. **Conservation:** This native species is relatively common and there are no national conservation measures at present. **Literature:** Bucke (1974), Cowx (1983), Goldspink (1981), Hartley (1947a,b), Kennedy & Fitzmaurice (1968a), Leeming (1964), Pitts et al. (1997), Svardson (1950), Wood & Jordan (1987), Wright (1990a).

Bleak *Alburnus alburnus* (Figs 18, 33; Plate 19)

Distinctive features: Body slim; base of anal fin longer than base of dorsal fin; more than 17 rays in the anal fin. 46–53 lateral cycloid scales. **Colour:**

Very silvery, the back is dark bluish green quickly grading to silver on the sides and silvery white on the belly. **Size:** 12–15 cm; maximum 25 cm; British rod record 129.4 g (1998: River Lark). **Age:** Average adult age 4–8 years. **Distribution:** Occurs throughout much of Europe from France east to the Caspian Sea. Indigenous to southeast England where it is common locally (Map p. 184). Introduced to Wales. Absent from Ireland. **Habitat:** Found in slow-flowing rivers and clear lakes. **Reproduction:** May–July among stones and gravel in shallow water. The adhesive yellow eggs (1.5 mm) hatch in 5–10 days and the young mature after 2–3 years. 5,000–6,500 eggs per female. Hybridises with Chub, Dace and Roach. **Food:** Invertebrates, especially crustaceans and insects (larvae and adults). **Value:** Important commercially only locally in parts of Europe, and may be used for animal food. Formerly important for its scales which were used as a material for covering artificial pearls. Of some sporting value in England and parts of continental Europe. **Conservation:** This native species, though local, is relatively common and there are no national conservation measures at present. **Literature:** Backe-Hansen (1982), Bastl (1977), Biro (1975), Harris & Wheeler (1974), Hartley (1947a,b), Politou et al. (1993), Spataru (1967b), Wheeler (1978a), Williams (1965, 1967).

Barbel *Barbus barbus* (Figs 18, 19; Plate 15)

Distinctive features: Inferior mouth with four sensory barbels; 55–65 lateral cycloid scales; last unbranched ray in the dorsal fin thickened and with numerous denticles posteriorly. **Colour:** A brownish green back, grading to golden brown on the sides and then to creamy white on the belly. **Size:** 25–75 cm; maximum 90 cm; British rod record 7.825 kg (1999: Great Ouse). **Age:** Average adult age 8–16 years. **Distribution:** Occurs across the middle of Europe from eastern England to the Black Sea. Indigenous only to southeast England but has been introduced to large rivers in other parts of England and to Wales, and recently to the River Clyde in Scotland (Map p. 184). Absent from Ireland. **Habitat:** Found mainly in the middle reaches of large rivers. **Reproduction:** May–July among gravel and stones in flowing water. The adhesive yellow eggs (2 mm) – which are reputed to be poisonous – hatch in 10–15 days and the young mature after 4–5 years. 3,000–32,000 eggs per female. **Food:** Invertebrates (mainly worms, molluscs and insect larvae) and some plant material when young; invertebrates and small fish when older. **Value:** Of some commercial importance locally in Europe, where it is caught by traps and nets. An important angling species in several countries, including

England. **Conservation:** This native species is relatively common and there are no national conservation measures at present. **Literature:** Baras & Philippart (1999), Hancock et al. (1976), Hunt & Jones (1974a,b, 1975), Poncin (1992), Tyler & Everett (1993), Wheeler & Jordan (1990).

Goldfish *Carassius auratus* (Figs 18, 21)

Distinctive features: Body of moderate depth and laterally compressed; 27–31 cycloid scales along the lateral line; dorsal fin concave, first ray strong and coarsely serrated. The Gibel Carp *Carassius auratus gibelio* is a common subspecies. **Colour:** Mainly dull greenish brown, lightening on the belly. However, in the domesticated forms, the dull young may change to one of the following colours or a combination of white, yellow, gold, red, blue or black, depending on the variety. **Size:** 15–35 cm; maximum 45 cm; maximum weight about 3 kg. British rod record 2.594 kg (1994: undisclosed water). **Age:** Average adult age 8–15 years. **Distribution:** Native to eastern Asia, this fish is now found in many places in Europe and other parts of the world to which it has been introduced. Established populations occur sporadically in various parts of Britain (Map p. 185). Absent from Ireland. **Habitat:** Occurs in rich weedy ponds, lakes and slow-flowing rivers. **Reproduction:** May–June, spawning among thick weed in shallow water. The adhesive yellow eggs (1.5 mm) hatch in 5–8 days and the young mature at 2–4 years. They may live for up to 20 years or more. 160,000–383,000 eggs per female. Hybridises with Crucian Carp. **Food:** Invertebrates and plant material. **Value:** Of some considerable commercial importance since the golden variety is one of the most popular aquarium fish, and large numbers are reared on fish farms in southern Europe for this purpose. Of little sporting value although occasionally used as a bait species. **Conservation:** As an alien species which is not threatened in areas where it is indigenous, no conservation measures are necessary in Britain. **Literature:** Allen (1987), Buth (1984), Davies (1963), Gerlach (1983), Hervey & Hems (1968), Lever (1977), Maitland (1971), Raicu et al. (1981), Seaman (1979), Street et al. (1984), Taylor & Mahon (1977).

Crucian Carp *Carassius carassius* (Figs 18, 22; Plate 8)

Distinctive features: Body usually deep and laterally compressed; 31–36 cycloid scales along the lateral line; dorsal fin convex, first ray feeble and weakly serrated. **Colour:** Olive-green on the back, grading to brassy green on the sides and dull brown on the belly. **Size:** 20–45 cm; maximum 50 cm; maximum weight 5 kg; British rod record 1.970 kg (1999: undisclosed water). **Age:** Average adult age 8–15 years. **Distribution:** Found throughout all of eastern and central Europe and many parts of the west. It is not native to many of these areas, but has been successfully introduced to many countries, including Britain where it is especially successful in the south of England (Map p. 185). Absent from Ireland. **Habitat:** Occurs in ponds, lakes and slow-flowing rivers. It is an exceptionally hardy species and can withstand very low temperatures and anoxic conditions for considerable periods. **Reproduction:** May–June, spawning in shallow water among thick weed growth. The adhesive yellow eggs (1.5 mm) hatch after 5–8 days and the young mature after 3–4 years. 137,000–244,000 eggs per female. Hybridises with Goldfish and Common Carp. **Food:** Bottom-dwelling invertebrates and plants. **Value:** Of limited commercial importance (mainly in fish farms in central Europe), but a useful sport species in some countries, including Britain. **Conservation:** As an alien species which is not threatened in areas where it is indigenous, no conservation measures are necessary in Britain. **Literature:** Adams & Mitchell (1992), Buth (1984), Marlborough (1966), Nilsson (1990), Raicu et al. (1981), Valente (1988).

Common Carp *Cyprinus carpio* (Figs 16, 18; Plate 17)

Distinctive features: Body covered with large scales; 33–40 cycloid scales along the lateral line; upper lip with two long barbels and two short barbels. **Colour:** The back is dull olive-brown, lightening to lighter brown on the sides and yellowish brown on the belly. **Size:** 25–75 cm; maximum 1.02 m; British rod record 25.572 kg (1998: Wraysbury). **Age:** Average adult age 10–30 years. **Distribution:** Native to eastern Europe and Asia, this fish has been successfully introduced to many other countries, including Ireland and Britain, where it is firmly established in many waters, especially in southern areas of England (Map p. 186). **Habitat:** It is found in rich, weedy ponds, lakes and slow-flowing rivers. **Reproduction:** June–July, spawning among thick weed in shallow water. The adhesive yellow eggs (1.5 mm) hatch in 3–6 days and the young mature after 3–5 years. 93,000–1,664,000 eggs per female. Hybridises with Crucian Carp. **Food:** Bottom-dwelling invertebrates and plant material. **Value:** Of considerable commercial value in many countries, particularly in central Europe, where many hundreds of fish farms are devoted to this species. Also of notable value across Europe as an elusive sport fish, when leather and mirror varieties are often encountered. The golden and coloured varieties are popular pond fish, especially the Koi varieties. **Conservation:** As an alien species which is not threatened in areas where it is indigenous, no conservation measures are necessary in Britain. **Literature:** Beukema & de Vos (1974), Beveridge et al. (1991), Crivelli (1981), Drost & Van den Boogaart (1986), Ede & Carlson (1977), Fitzmaurice (1983a), Garcia & Martin (1985), Gerlach (1983), Heap & Goldspink (1986), Korwin-Kossakowski (1988), Leeming (1970), O'Grady & Spillett (1985), Pottinger (1998), Soto et al. (1994), Stein & Kitchell (1975) Taylor & Mahon (1977), Vilizzi (1998).

Common Gudgeon *Gobio gobio* (Figs 18, 20; Plate 16)

Distinctive features: Two well-developed barbels; scales large and cycloid, 40–45 along the lateral line. **Colour:** The back is olive-brown with some dark spotting, grading to lighter brown on the sides and dull whitish yellow on the belly. **Size:** 10–15 cm; maximum 20 cm; British rod record 141.8 g (1989: River Nadder). **Age:** Average adult age 5–8 years. **Distribution:** Found throughout temperate Europe and Asia. Indigenous to England but introduced to Ireland, Wales and parts of Scotland (Map p. 186). **Habitat:** Occurs in rich sandy streams and rivers, and some lakes and canals. **Reproduction:** May–June, among stones and weed in running water. The young hatch after 15–20 days and adults mature after 2–3 years, living up to a maximum of 8 years. 800–3,000 adhesive yellow eggs (1.8 mm) per female. **Food:** Invertebrates, especially molluscs, crustaceans and insect larvae. Occasionally plant material. **Value:** Of little commercial significance, but occasionally used as a bait species in sport fishing, or angled for in its own right. **Conservation:** This native species is relatively common and there are no national conservation measures at present. **Literature**: Bean & Winfield (1989, 1992), Coelho (1981), Hartley (1947a), Kennedy & Fitzmaurice (1972a), Ladich (1988), Mann (1980), Radforth (1940), Stott (1967), Stott et al. (1963).

Sunbleak *Leucaspius delineatus* (Figs 4, 32; Plate 12)

Distinctive features: Lateral line incomplete, extending over the first 2–13 scales (cycloid) only; ventral fin with 10 rays. Also known as Motherless Minnow or Belica. **Colour:** The back is olive-green, changing to metallic silver on the sides and belly. **Size:** 5–9 cm; maximum 12 cm. **Age:** Average adult age 4–6 years. **Distribution:** Occurs in central and eastern Europe from the Rhine basin in the west to the Volga basin in the east. Introduced in recent years to England where it is successfully established in several waters (Map p. 187), for example Sedgemoor Drain, Somerset, where it was first discovered in March 1990, and in a fish farm at Skegness, Lincolnshire, where it was found in June 1991. **Habitat:** Found in small ponds, and the lower reaches of some lowland streams and rivers **Reproduction:** June–July, spawning among weeds in shallow water. Matures after 2 years and lives for only about 5 years. **Food:** Invertebrates (especially crustaceans and some insect larvae) and fine plant material. **Value:** Of local commercial importance in net fisheries in parts of Russia. Rarely angled for but occasionally kept in aquaria. **Conservation:** As an alien species which is not threatened in areas where it is indigenous, no conservation measures are necessary in Britain. **Literature:** Farr-Cox et al. (1996), Maitland (2000).

Chub *Leuciscus cephalus* (Figs 18, 28; Plate 20)

Distinctive features: Body with large silvery cycloid scales, less than 46 along the lateral line; forehead wide and flat; anal fin rounded. **Colour:** Dark olive greenish grey on the back, grading to greenish silver on the sides and yellowish white on the belly. **Size:** 30–50 cm; maximum 80 cm; British rod record 3.912 kg (1994: River Tees). **Age:** Average adult age 5–10 years. **Distribution:** Found throughout most of central and southern Europe, sometimes in brackish water (e.g. Baltic Sea). In Britain it is indigenous only to England but has been introduced to Wales and southern parts of Scotland (Map p. 187). Absent from Ireland. **Habitat:** Occurs in the middle reaches of rivers and occasionally in lakes. **Reproduction:** April–June among stones and plants in slow-flowing water. The adhesive yellow eggs (2 mm) hatch in 6–8 days and the adults are mature after 3–4 years. 50,000–200,000 eggs per female. Hybridises with Bleak. **Food:** Mainly invertebrates and some plant material when young, large invertebrates, sometimes fruit, and fish when adult. **Value:** Only locally of importance in parts of Europe as a commercial species, where it is caught in nets. It is of considerable sporting value in many parts of Europe, including Britain. **Conservation:** This native species is relatively common and there are no national conservation measures at present. **Literature:** Adams et al. (1990), Cragg-Hine & Jones (1969), Hellawell (1971b,c,d), Hickley & Bailey (1982), Leeming (1964), Mann (1976b), Penaz (1968), Wheeler & Easton (1978).

Orfe *Leuciscus idus* (Figs 18, 25, 27; Plate 21)

Distinctive features: Body with well-developed cycloid scales, 55–61 along the lateral line; 12–14 rays in the dorsal fin. Also known as Ide. **Colour:** A dark brownish grey back, grading to silvery grey on the sides and silver on the belly. **Size:** 25–50 cm; maximum 100 cm; British rod record 3.778 kg (2000: Lymm Vale). **Age:** Average adult age 5–10 years. **Distribution:** Native to most of Europe and parts of Asia east of the Rhine. Introduced successfully to several countries west of the Rhine, including Britain where it is naturalised in several waters, especially in parts of England (Map p. 188). Absent from Ireland. **Habitat:** Found sometimes in brackish water, it occurs mainly in rivers and some lakes. **Reproduction:** April–May, in shallow water among weed and stones. The adhesive yellow eggs (1.5 mm) hatch in 15–20 days and the adults mature at 5–7 years. 39,000–114,000 eggs per female. Hybridises with Common Bream. **Food:** Invertebrates, especially molluscs, crustaceans and insect larvae, and some plant material, when young; invertebrates and some fish when larger. **Value:** Of considerable commercial importance in Russia, where large numbers are taken in nets and traps. Angled for in some parts of Europe. The golden variety is very popular as an aquarium or pond fish. **Conservation:** As an alien species which is not threatened in areas where it is indigenous, no conservation measures are necessary in Britain. **Literature:** Cala (1971), Lever (1977).

Dace *Leuciscus leuciscus* (Figs 18, 29; Plate 22)

Distinctive features: Body with large silvery cycloid scales, 49–52 along the lateral line; anal fin concave; 10–11 rays in the dorsal fin. **Colour:** Greenish grey on the back, grading to silvery grey on the sides and silvery white on the belly. **Size:** 15–25 cm; maximum 30 cm; British rod record 567 g (1960: River Little Ouse). **Age:** Average adult age 4–8 years. **Distribution:** Found throughout Europe except the extreme northwest. Indigenous to southeast England, but has been introduced to other regions including southern Ireland, Wales and southern Scotland (Map p. 188). **Habitat:** Occurs in clear rivers and streams Occasionally in lakes or in brackish water near river mouths. **Reproduction:** March–May, among stones and plants in running water. The adults mature after 3–4 years. 2,500–27,500 adhesive yellow eggs (1.5 mm) per female. **Food:** Mainly invertebrates, especially aquatic insects – both larvae and adults. Hybridises with Bleak and Rudd. **Value:** Of some commercial importance in parts of Russia. A sport species in some parts of western Europe, including Britain. **Conservation:** This native species is relatively common and spreading north through introductions by anglers. There are no national conservation measures at present. **Literature:** Adams et al. (1990), Clough et al. (1998), Cowx (1988), Cragg-Hine & Jones (1969), Hartley (1947a), Hellawell (1974), Hickley & Bailey (1982), Kennedy (1969), Kennedy & Hine (1969), Mann (1974), Mann & Mills (1983), Mathews & Williams (1972), Mills (1980, 1982), Weatherley (1987), Williams (1967), Winfield et al. (2004).

Common Minnow *Phoxinus phoxinus* (Figs 18, 25, 26; Plate 11)

Distinctive features: Scales (cycloid) small; 80–100 along the lateral line; numerous brown and black blotches along the sides, sometimes uniting to form stripes; males brightly coloured during spawning. **Colour:** Very variable, depending on environment and season. Normally brownish green on the back, separated from whitish grey on the belly by a longitudinal dark line or series of blotches. The breeding dress of the male has white flashes at the fins, reddish pectoral and pelvic fins, a black throat, green along the sides and a scarlet belly. **Size:** 6–10 cm; maximum 14 cm. British rod record 135 g (1998: Whitworth Lake). **Age:** Average adult age 3–5 years. **Distribution:** Found over almost the whole of Europe and northern Asia. Indigenous to much of the British Isles, except the extreme northwest of Scotland and most northern islands (Map p. 189). **Habitat:** Occurs in streams and rivers, ponds and lakes **Reproduction:** June–July, spawning in shoals over stones and gravel (to which the eggs adhere) in running water. The adhesive yellow eggs (1.5 mm) hatch in 5–10 days and the adults mature after 2–3 years. 200–1,000 eggs per female. **Food:** Invertebrates, especially crustaceans and insect larvae, and some plant material. **Value:** Of commercial importance in parts of Russia. A valuable bait species elsewhere. Commonly used also as an aquarium and laboratory fish. **Conservation:** As this native species is very common there are no national conservation measures at present. **Literature:** Bibby (1972), Frost (1943), Garner et al. (1998), Greenwood & Metcalfe (1998), Griffiths (1997), Lein (1981), Levesley & Magurran (1988), Magurran (1986), Maitland (1965), Mills (1988), Naish et al. (1993), Pitcher et al. (1986), Rasottos et al. (1987), Stott & Buckley (1979), Wootton & Mills (1979).

False Harlequin *Pseudorasbora parva* (Fig. 23; Plate 13)

Distinctive features: Mouth slightly superior and oblique; lateral line straight; 34–38 lateral scales; narrow dark line from head to tail. Also known as

Topmouth Gudgeon. **Colour:** Bluish grey on the back, changing to silvery on the sides and white on the belly. A darkish lateral stripe runs from head to tail. **Size:** 4–6 cm; maximum 7 cm. **Age:** Average adult age 2–3 years. **Distribution:** Introduced to Europe from Asia, where it is native to the Amur River and other eastern catchments. Now introduced to many parts of Europe, including England where it is well established in at least three different catchments (Map p. 189). **Habitat:** Slow-flowing rivers and lakes. **Reproduction:** Spawns during summer at 21–26°C. Batch spawning with some guarding of the eggs by parents. **Food:** Invertebrates, especially planktonic crustaceans, and fish fry. **Value:** Of little commercial or angling importance, though it has been intentionally stocked in some waters in Europe as a forage food for piscivorous fish. Sometimes kept by aquarists. **Conservation:** As an alien species which is not threatened in areas where it is indigenous, no conservation measures are necessary in Britain. **Literature:** Allardi & Chancerel (1988), Bianco (1988), Gozlan et al. (2002), Knezevic (1981), Rosecchi et al. (2001), Xie et al. (2001).

Bitterling *Rhodeus sericeus* (Figs 18, 24; Plate 14)

Distinctive features: Body deep and compressed laterally; scales large and cycloid, 32–40 laterally; lateral line incomplete; female bears a long ovipositor. **Colour:** The back is greenish grey, grading to bluish silver on the sides and silvery white on the belly. An iridescent blue stripe runs along the sides, sometimes with a reddish edge. **Size:** 5–8 cm; maximum 10 cm. **Age:** Average adult age 3–5 years. British rod record 37.2 g (1998: Barway Lake). **Distribution:** Occurs throughout much of the middle of Europe from France eastward to the Caspian Sea. Introduced successfully to a number of countries, including England, where it is common in the Cheshire area and is gradually spreading east through local canal systems (Map p. 190). **Habitat:** Found in rich, sandy, slow-flowing rivers and lakes, where mussels are present. **Reproduction:** April–June, the fish form pairs and the female deposits eggs into the mantle cavity of large freshwater unionid mussels, by means of an elongate ovipositor. The adhesive yellow eggs (3 mm) hatch in 15–20 days and the young leave the mussel a few days later. The young mature after 2–3 years and may live for 5 or more years. 40–100 eggs per female. **Food:** Both plants (mainly filamentous and other attached algae) and invertebrates (especially insect larvae and small crustaceans) are eaten. **Value:** Of little commercial value, but often used as a bait fish and very commonly kept in aquaria because of its interesting breeding habits. **Conservation:** Though it is listed in Appendix III of the Bern Convention, as an alien species which is not threatened in areas where it is indigenous, no conservation measures are necessary in Britain. **Literature:** Aldridge (1999), De Wit (1954), Hardy (1954), Papadopol (1960), Spataru & Gruia (1967), Wheeler & Maitland (1973).

Roach *Rutilus rutilus* (Figs 18, 30; Plate 9)

Distinctive features: Body with large silvery cycloid scales; pectoral fins with 16 rays. **Colour:** Greenish brown on the back, grading to silvery brown on the sides and pale white on the belly. **Size:** 20–35 cm; maximum 44 cm – this fish weighed 2.1 kg; British rod record 1.899 kg (1990: River Stour). **Age:** Average adult age 4–8 years. **Distribution:** Occurs over most of northern Europe and Asia. Sometimes found in brackish water near the mouths of large rivers, e.g. in the Baltic and Black Seas. It is common also in parts of the Caspian Sea. Indigenous to much of England, Wales and Scotland where it is established as far north as the Great Glen (Map p. 190). Introduced to Ireland. **Habitat:** Common in rich ponds, lakes, canals and slow-flowing rivers. **Reproduction:** May–June, among weed in shallow water, usually in standing water but sometimes moving into fast-flowing water to spawn. The adhesive yellow eggs (1.4 mm) hatch in 5–10 days and the larvae move around in large shoals, as do the adults which mature in about 2–3 years. 1,000–14,600 eggs per female. Hybridises with Silver Bream, Common Bream, Bleak and Rudd. **Food:** An omnivorous species, feeding on both aquatic plants (especially attached algae) and invertebrates of many kinds. **Value:** Commercially important in some parts of eastern Europe where it is caught in traps and nets. An important sport species in many countries, including Britain. Small roach are sometimes used as bait in pike fishing. **Conservation:** This native species is relatively common and there are no national conservation measures at present. **Literature:** Ali (1976, 1979), Broughton & Jones (1978), Burrough (1978), Burrough & Kennedy (1979), Child & Solomon (1977), Copp (1990), Cragg-Hine & Jones (1969), Goldspink (1978), Hartley (1947a), Hellawell (1972), Jones (1953), Krause et al. (1998), Linfield (1979), Mann (1973), Mills (1969), Radforth (1940), Sweeting (1976), Townsend & Perrow (1989), Treasurer (1990b), Vollestad & L'Abee-Lund (1990), Wheeler & Easton (1978), Williams (1965, 1967), Winfield et al. (1992), Wood & Jordan (1987), Wyatt (1988).

Rudd *Scardinius erythrophthalmus* (Figs 18, 31; Plate 10)

Distinctive features: Body with large silvery cycloid scales, 40–43 along the lateral line; lower fins bright red; origin of pelvic fins anterior to that of the dorsal fin. Some workers believe that this species should actually be placed in the genus *Rutilus*. **Colour:** Dark olive-brown on the back, grading to yellowish silver on the sides and dull white on the belly. **Size:** 15–30 cm; maximum 45 cm; British rod record 2.041 kg (1933: undisclosed mere). **Age:** Average adult age 4–8 years. **Distribution:** Found throughout much of Europe, except the extreme north and west. Native to England but introduced to Ireland, Wales and a few places in southern Scotland (Map p. 191). **Habitat:** Occurs in rich, weedy ponds, lakes, canals and slow-flowing rivers **Reproduction:** May–June, spawning in shoals among weed. The adhesive yellow eggs (1.1 mm) hatch in 5–8 days and the adults mature after 2–3 years. 90,000–200,000 eggs per female. Hybridises with Silver Bream, Common Bream and Roach. **Food:** Invertebrates (especially molluscs and insect larvae) and aquatic vegetation. **Value:** Of little commercial value, but of value as a sport and a bait species in some countries, including England. The golden variety is commonly kept in aquaria and ponds. **Conservation:** This native species is relatively common and there are no national conservation measures at present. **Literature:** Bean & Winfield (1995), Brassington & Ferguson (1975), Burrough (1978), Hartley (1947a,b), Johansson (1987), Kennedy & Fitzmaurice (1974), Mann & Steinmetz (1985), Svardson (1950), Thompson & Iliadou (1990), Wheeler (1976), Winfield (1986).

Tench *Tinca tinca* (Figs 3, 17, 18; Plate 18)

Distinctive features: Body deep, with small, deeply embedded cycloid scales, 95–120 along the lateral line; one pair of small barbels at the mouth; fins very rounded. **Colour:** Dark olive-brown on the back, grading to greenish brown on the sides and dull creamy brown on the belly. **Size:** 20–40 cm; maximum 65 cm; British rod record 7.910 kg (1993: undisclosed water). **Age:** Average adult age 10–20 years. **Distribution:** Occurs over most of Europe except northern areas. Indigenous to England, but it has been introduced successfully to Ireland, Wales and southern Scotland (Map p. 191). **Habitat:** Found in rich, weedy lakes and also in slow-flowing rivers.

Reproduction: May–July, spawning among weed in shallow water. The adhesive yellow eggs (1.1 mm) hatch in 3–5 days, and adults mature after 3–4 years. They may live for up to 10 years. 280,000–827,000 eggs per female. **Food:** Invertebrates, especially molluscs, crustaceans and insect larvae. **Value:** Of relatively little commercial importance, except in central Europe where it is farmed for the table. Valued as a sport species in many countries, including Britain and Ireland, where it is caught with baited hooks of various types. The golden variety is a popular aquarium and pond fish. **Conservation:** This native species is relatively common and there are no national conservation measures at present. **Literature:** Giles et al. (1990), Kennedy & Fitzmaurice (1970), Petridis (1990), Rosa (1958), Weatherley (1959, 1962).

SPINED LOACH FAMILY: COBITIDAE

Spined Loach *Cobitis taenia* (Figs 36, 37; Plate 23)

Distinctive features: Six barbels around the mouth, and a movable double spine below each eye; head compressed laterally; male pectoral fins with a thickened second ray. Scales cycloid and very small. **Colour:** The back is pale brown with dark brown to black marbling. There are two lateral rows of dark brown spots along the pale brown sides. The belly is silvery white. **Size:** 8–10 cm; maximum 14 cm. **Age:** Average adult age 3–5 years. **Distribution:** Found over most of Europe except the extreme north and west. Native to southeast England and absent so far from other parts of the British Isles (Map p. 192). **Habitat:** Occurs mainly in slow-flowing streams and lakes. **Reproduction:** April–July, laying yellow eggs (0.8 mm) among stones and weed in shallow, often silty, running waters. The larvae become bottom-living almost immediately after hatching. **Food:** Bottom-dwelling invertebrates, especially worms, small molluscs and insect larvae. **Value:** Of no commercial or sporting value. **Conservation:** This native species is listed in Annex IIa of the EU Habitats and Species Directive and in Appendix III of the Bern Convention. Though very local in its distribution, it is relatively common where it occurs and there are no national conservation measures at present. **Literature:** Bohlen (1998, 1999), Lodi (1967, 1980), Robotham (1977, 1982a,b), Spataru (1967a).

STONE LOACH FAMILY: BALITORIDAE

Stone Loach *Barbatula barbatula* (Figs 36, 38; Plate 24)

Distinctive features: Six barbels around the mouth but no spine under the eye; head rounded; caudal fin truncated. Scales cycloid and very small. Formerly *Noemacheilus barbatulus*. **Colour:** Variable, but mostly dark olive-brown on the back, grading through yellow-grey on the sides to whitish on the belly. The back and sides usually have an irregular dark mottling. **Size:** 8–12 cm; maximum 18 cm. **Age:** Average adult age 4–6 years. **Distribution:** Found throughout much of Europe except the extreme north and south. Indigenous to Britain but absent from the north of Scotland (except Orkney where a recently introduced stock is now established) (Map p. 192). **Habitat:** Occurs mainly in stony streams and rivers but is also found in oligotrophic lakes **Reproduction:** April–May, laying yellow eggs (0.9 mm) among stones and weed in running water. The young mature after 2–3 years and may live for up to 8 years. 500,000–800,000 eggs per female. **Food:** Benthic invertebrates, especially insect larvae of various kinds. **Value:** Occasionally eaten, but not of real commercial use. Sometimes used in sport fishing as a bait species. **Conservation:** This native species is relatively common and there are no national conservation measures at present. **Literature:** Hyslop (1982), Maitland (1965), Mills et al. (1983), Rumpus (1975), Skryabin (1993), Smyly (1955), Street & Hart (1985).

AMERICAN CATFISH FAMILY: ICTALURIDAE

Black Bullhead *Ameiurus melas* (Fig. 39; Plate 25)

Distinctive features: Broad flat head with eight barbels; barbs on posterior pectoral spines weak or absent; anal fin with 17–21 rays. No scales. **Colour:** The back is dark brownish black, changing to lustrous greenish gold on the sides and muddy white on the belly. Spawning males are jet black, with a yellowish white belly. **Size:** 20–30 cm; maximum 45 cm; maximum weight about 3 kg. British rod record 455.4 g (1999: Lake Meadows). **Age:** Average adult age 6–12 years. **Distribution:** Native to eastern North America, but introduced to Europe where it has established itself successfully in several countries, including England, where it occurs in a few waters (Map p. 193).

Habitat: Occurs in rich slow-flowing rivers and lakes. **Reproduction:** June–July, in a nest among logs, plants etc. Eggs and young are guarded by both parents. 2,550–3,850 eggs per female. Matures at 2–3 years and may spawn annually for about 9 years. **Food:** Both plants and invertebrates, especially crustaceans. Small fish are also eaten. **Value:** Of little commercial or sporting value in Europe but often sold for aquaria and ponds. Of considerable value in North America for angling and pond culture. **Conservation:** As an alien species which is not threatened in areas where it is indigenous, no conservation measures are necessary in Britain. **Literature:** Becker (1983), Campbell & Branson (1978), Carlander (1977), Cloutman (1979), Scott & Crossman (1973), Tin (1982a), Trautman (1981), Wallace (1967), Wheeler (1978b).

EUROPEAN CATFISH FAMILY: SILURIDAE

Wels Catfish *Silurus glanis* (Fig. 40; Plate 26)

Distinctive features: Large flat head with two long barbels on the upper jaw; long anal fin. No scales. Also known as Danube Catfish. **Colour:** Variable in colour; the back is normally dark bluish or brownish black, sometimes with a reddish sheen. The sides are pale with dark marbling and spotting. **Size:** 1–2 m; maximum 5 m – this fish weighed 306 kg; British rod record 28.123 kg (1997: Withy Pool). **Age:** Average adult age 15–25 years. **Distribution:** Occurs in suitable waters across most of central and eastern Europe. Also in brackish water in the Baltic and Black Seas. Introduced successfully to a number of other countries including England, where it is being spread by anglers from a few waters in the southeast to other catchments in the west and north (Map p. 193). Absent from Ireland. **Habitat:** Found in the lower reaches of large rivers and in muddy lakes. **Reproduction:** May–July, spawning in shallow water under thick vegetation where the male fish excavates a depression in which the adhesive yellow eggs (3 mm) are laid. These are guarded by the male until they hatch in 2–3 days. The young mature after 4–5 years and may live for up to 15 years. 136,000–467,000 eggs per female. **Food:** Invertebrates when young, vertebrates (especially fish but also amphibians and waterfowl) when older. **Value:** Of considerable commercial importance in eastern Europe where it is caught in nets, in traps and on large baited hooks. It is also produced in a few fish farms. Eggs may be used as caviar. Prized as a large angling species in some areas, including England, where it is caught on set lines baited with fish or frogs. **Conservation:** Though listed in Appendix III of the Bern Convention, as an alien species which is not especially threatened in areas where it is indigenous no conservation measures are necessary in Britain. **Literature:** Krasznai & Marian (1986), Lever (1977), Maitland (1983), Schindler (1957), Wheeler & Maitland (1973).

PIKE FAMILY: ESOCIDAE

Pike *Esox lucius* (Fig. 41; Plate 36)

Distinctive features: Head and snout large, with enormous mouth and teeth. Slim streamlined body with dorsal and anal fins well back towards the tail; 110–130 lateral cycloid scales. **Colour:** The back is green to olive-brown, grading on the flanks to pale green with oval yellow patches. The belly is pale creamy yellow, sometimes with pale orange spots. **Size:** 30–120 cm; maximum 150 cm; weight up to 34 kg; British rod record 21.234 kg (1992: Llandegfedd). **Age:** Average adult age 10–15 years. **Distribution:** Common across most of temperate Europe, Asia and North America, including the British Isles where it is widespread, except in the extreme north and west, and most islands (Map p. 194). **Habitat:** Occurs in both standing and slow-flowing waters, especially where weed cover is present. **Reproduction:** February–May among weed. The adhesive dark yellow eggs (2.7 mm) hatch after 10–15 days and larvae attach to weed for a short time. They mature in 3–4 years and may live for 25 years. 16,600–79,700 eggs per female. **Food:** Invertebrates at first, but soon fish and other vertebrates are eaten. Cannibalistic and can exist on its own kind where no other species is present. **Value:** Of commercial importance and a popular table fish in central and other parts of Europe. A major sport species across its distribution, including Britain and Ireland. **Conservation:** This native species is very common and there are no national conservation measures at present. **Literature:** Beukema (1970), Bregazzi & Kennedy (1980), Craig & Kipling (1983), Fabricius & Gustavson (1958), Fickling (1982), Fitzmaurice (1983b), Healy (1956), Healy & Mulcahy (1980), Kipling & Frost (1969), Mann (1976a, 1982), Munro (1957), Salam & Davies (1994), Shafi & Maitland (1971a), Svardson (1950), Toner (1959), Treasurer (1990a), Wright (1990b).

SMELT FAMILY: OSMERIDAE

Smelt *Osmerus eperlanus* (Fig. 42)

Distinctive features: Small silvery fish with an adipose fin; pelvic fins without axillary processes; large mouth and teeth; lateral line incomplete. 60–66 lateral cycloid scales. Also known as Sparling in northern England and Scotland. **Colour:** The back is dark grey-green, shading to silvery grey along the sides and silvery white on the belly. **Size:** 10–20 cm; maximum 31 cm; fish 16 cm long weigh about 32 g. British rod record 191 g (1981: Fleetwood, Lancashire). **Age:** Average adult age 2–5 years. **Distribution:** Occurs all round the coast of northwestern Europe. Purely freshwater populations occur in some Scandinavian lakes. Native to much of the British Isles, south of the midland valley of Scotland, though many stocks have declined due to river barriers and pollution (Map p. 194). **Habitat:** Found mainly in estuaries and lower reaches of accessible unpolluted rivers. **Reproduction:** March–April, spawning among weeds in rivers above estuaries, or at the edges of lakes. The pale yellow eggs (0.9 mm) hatch in 20–35 days and the young move down to the sea, maturing after 2 years. They live for up to 8 years. 10,000–40,000 eggs per female. **Food:** Invertebrates (especially crustaceans) when young, invertebrates and fish when older. **Value:** Of commercial value in some rivers where it is caught in nets and traps. Introduced as a forage fish to lakes in Scandinavia. **Conservation:** This native species has undergone significant decline in many parts of the British Isles and is in need of conservation action. Unfortunately, at present it does not receive any national legislative protection. **Literature:** Anders & Weise (1993), Banks (1970), Belyanina (1969), De Groot (1989), Ellison & Chubb (1968), Garnas (1982), Hutchinson (1983), Hutchinson & Mills (1987), Lees & Whitfield (1992), Lyle & Maitland (1996), Maitland & Lyle (1996), Masterman (1913), Naesje et al. (1987), Nellbring (1989).

WHITEFISH FAMILY: COREGONIDAE

Vendace *Coregonus albula* (Fig. 45; Plate 29)

Distinctive features: Mouth superior; 36–52 gill rakers; 75–88 lateral cycloid scales; usually small. Also known as Cisco. **Colour:** The back is dark greenish blue, grading to silvery green along the sides and silvery white on the belly. **Size:** 15–25 cm; maximum 33 cm; fish with a mean weight of 18 cm weigh

60 g. **Age:** Average adult age 3–7 years. **Distribution:** Found in many parts of northern and central Europe. Some populations in the Baltic are anadromous. Only four populations have been recorded from Britain: two in Scotland (Castle and Mill Lochs, Lochmaben), both of which are now extinct, and two in England (Bassenthwaite Lake and Derwent Water, Cumbria); recently re-introduced successfully to Scotland (Loch Skene) (Map p. 195). **Habitat:** Usually in deep, cold oligotrophic and mesotrophic lakes. **Reproduction:** November–December; yellow eggs (2.1 mm) are laid over and sink among gravel or stones at depths of 2–3 m or more. Eggs hatch after 100–120 days and the young move into deeper water. Adults mature after 2–3 years and may live for up to 10 years. 1,700–4,800 eggs per female. **Food:** Invertebrates, especially planktonic crustaceans. **Value:** Of no sporting importance, but important commercially in several parts of Europe (e.g. Finland and Russia), fish being caught in nets and traps during their spawning migration. Sometimes smoked. **Conservation:** Because of its decline, the Vendace is now given some protection. It is listed in Schedule 5 of the 1981 Wildlife and Countryside Act, Annex Va of the EU Habitats and Species Directive and in Appendix III of the Bern Convention, and as a Priority Species in the UK Biodiversity Action Plan. There is no Red Data Book for fish in Britain, but Maitland (2000) considers this species to be Vulnerable. **Literature:** Aass (1972), Beaumont et al. (1995), Dembinski (1971), Enderlein (1989), Ferguson (1974), Jurvelius et al. (1988), Maitland (1966a,b, 1967b, 1970, 1982), Sutela & Huusko (1997), Vuorinen et al. (1991), Winfield et al. (1996, 2004).

Pollan *Coregonus autumnalis* (Fig. 44; Plate 27)

Distinctive features: Mouth terminal, 74–86 lateral cycloid scales, 35–51 gill rakers. Also known as Arctic Cisco. **Colour:** The back is dark greenish brown grading to silvery green along the sides and silvery white on the belly. **Size:** 30–35 cm, 450 gm; maximum 38 cm, exceptionally 45 cm (1.3 kg) in Canada and even 50 cm (2 kg) in Siberia. **Age:** Average adult age 4–9 years. **Distribution:** Occurs in western Europe only in Ireland (e.g. Lough Neagh) (Map p. 195); also found in northern Eurasia and North America, where it is anadromous. **Habitat:** Large lakes in Ireland, but also coastal waters and the lower parts of Arctic rivers elsewhere. **Reproduction:** October–December over areas of gravel and stones. Yellowish eggs (2.5 mm) hatch in March and the young reach 20 cm in 2 years and are adult at 3–4 years. They normally live for 5–7 years (maximum 9–10). 2,000–8,000 eggs per female (exceptionally 90,000 in very large females). **Food:** Zooplankton when young, zooplankton and bottom invertebrates when older. **Value:** Fished commercially in Ireland using seine and gill nets. Important commercially in both Siberia and North America. **Conservation:** Protected by its listing in Annex Va of the Habitats and Species Directive, and Appendix III of the Bern Convention, it is listed as Endangered in the Irish Red Data Book of Vertebrates (Whilde 1993). **Literature:** Dabrowski & Jewson (1984), Fechhelm et al. (1993), Ferguson (1974), Ferguson et al. (1978), Harrod et al. (2001), Maitland (1970), McPhail (1966), Twomey (1956), Wilson (1983, 1984), Wilson & Pitcher (1983, 1984), Winfield & Wood (1990), Winfield et al. (1993b).

Powan *Coregonus lavaretus* (Fig. 46; Plate 30)

Distinctive features: Mouth small, toothless and inferior; 17–48 gill rakers; 84–100 lateral cycloid scales. A very variable species forming local races in different lakes, many of them named as subspecies. Known in England as Schelly, in Wales as Gwyniad and in Scotland as Powan. **Colour:** The back is a dark bluish grey, grading to greenish grey along the sides to silvery whitish yellow on the belly. **Size:** 15–40 cm; maximum 57 cm; maximum weight about 2.8 kg; former British rod record 951.5 g (1986: Haweswater – note this is now a protected species). **Age:** Average adult age 4–9 years. **Distribution:** Northwestern and central Europe and in the Baltic area where it migrates into rivers to spawn. It occurs in only seven waters in Britain: two in Scotland (Lochs Lomond and Eck), four in England (Ullswater, Haweswater, Red Tarn and Brotherswater) and one in Wales (Llyn Tegid) (Map p. 196). **Habitat:** Normally in large, cold oligotrophic lakes. **Reproduction:** December–February, over gravel; timing may be influenced by moon phases. Yellowish eggs (2.5 mm) hatch in about 100 days. Young mature in 3–4 years and live for up to 10 years. 1,000–28,000 eggs per female. **Food:** Invertebrates and planktonic crustaceans when young, but both bottom-dwelling and planktonic forms later. **Value:** Of commercial importance in various countries in Europe, and caught mainly in gill nets. Occasionally angled for. **Conservation:** Because of threats to the few existing populations, *Coregonus lavaretus* is now given some protection. It is listed in Schedule 5 of the 1981 Wildlife and Countryside Act, in Annex Va of the EU Habitats and Species Directive and in Appendix III of the Bern Convention. There is no Red Data Book for fish in Britain, but Maitland (2000) considers this species to be Vulnerable. **Literature:** Adams & Tippett (1991), Ausen (1976), Bagenal (1970),

Beaumont et al. (1995), Brown (1989), Brown & Scott (1987), Dabrowski et al. (1984), Ellison (1966), Ellison & Cooper (1967), Fuller & Scott (1976), Gervers (1954), Haram (1968), Haram & Jones (1971), Maitland (1967a, 1969b, 1970, 1980b, 1982), Roberts et al. (1970), Scott (1975), Slack et al. (1957), Winfield et al. (1993a, 1995, 1996, 1998, 2002b, 2003).

Houting *Coregonus oxyrinchus* (Fig. 43)

Distinctive features: Mouth small, toothless and inferior; well-developed snout; 35–44 gill rakers. 80–88 lateral cycloid scales. Some believe this fish is merely a subspecies of *Coregonus lavaretus*. **Colour:** The back is dark bluish green, grading to silvery grey along the sides and creamy white on the belly. **Size:** 25–40 cm; maximum 50 cm; maximum weight about 2 kg. **Age:** Average adult age 4–9 years. **Distribution:** Northern Europe, especially the Baltic Sea, where large populations formerly occurred. Known in the British Isles only from specimens formerly caught off the coast of southwest England, where it no longer occurs. **Habitat:** Coastal areas in brackish water, only running into fresh water to spawn. **Reproduction:** October–December, in the lower reaches of rivers where the yellowish eggs (2.9 mm) are laid among stones and gravel. After hatching, the young migrate to the sea, where they mature in 3–4 years. **Food:** Invertebrates, planktonic crustaceans when young, planktonic and benthic forms later. **Value:** Formerly locally important in some areas and caught during the autumn spawning migration. Many populations are now extinct but recently a conservation programme in Denmark has proved successful in restoring some stocks. **Conservation:** This species is listed in Appendix III of the Bern Convention but has not been seen in British waters for many decades and there is no special protection or conservation action for it here. However, very successful restoration programmes are being carried out in Denmark and the Netherlands, and the Houting may well appear again along British coasts. **Literature:** Berg (1965), Kranenbarg et al. (2002), Maitland (1970), Svardson (1956).

SALMON FAMILY: SALMONIDAE

Rainbow Trout *Oncorhynchus mykiss* (Figs 47, 50; Plate 31)

Distinctive features: Caudal and adipose fins heavily spotted; a broad pink band is usually present along the sides. 135–150 lateral cycloid scales.

Formerly *Salmo gairdneri*. **Colour:** The colour is variable and may range from very silvery to grey-brown on the back, grading to silvery brown on the sides and white on the belly. Superimposed on this there is usually a characteristic pink lateral band and dense black spotting – notable on the adipose and tail fins. **Size:** 25–45 cm; maximum 70 cm and 20 kg; British rod record 10.921 kg (1960: Hanningfield Reservoir). **Age:** Average adult age 3–6 years. **Distribution:** Native to the Pacific coastal basins of North America and Asia where both migratory (Steelhead Trout) and purely freshwater (Rainbow Trout) populations occur. Has been introduced widely all over Europe, although self-maintaining populations have developed in only a few places. In the British Isles there are very few such populations, though the species is stocked widely each year (Map p. 196). **Habitat:** Normally, in its native area, clear lakes, streams and rivers, but in the British Isles mainly lakes and reservoirs where it is introduced regularly for put-and-take fisheries. Occurs here in rivers mainly as escapes from fish farms. **Reproduction:** October–March, the eggs being laid in nests dug out of gravel in running water. The yellowish orange eggs (4 mm) hatch after 100–150 days and the young gradually move into larger rivers and lakes. 1,000–5,000 eggs per female. **Food:** Invertebrates when young, invertebrates and fish when adult. **Value:** Of major importance as a commercial species in many fish farms in Europe and North America, and as a stocked sporting species in rivers and lakes. Especially popular with anglers as a put-and-take fish in Britain and Ireland. **Conservation:** As an alien species which is not threatened in areas where it is indigenous, no conservation measures are necessary in Britain. **Literature:** Behnke (1984), Dodge & MacCrimmon (1970), Frost (1940, 1974), Hunt & O'Hara (1973), Kennedy & Strange (1978), Lever (1977), MacCrimmon (1971), Martinez (1984), Narver (1969), Phillips (1989), Scott & Crossman (1973), Smith & Stearley (1989), Withler (1966), Worthington (1941).

Atlantic Salmon *Salmo salar* (Figs 47, 51, 52)

Distinctive features: 10 or fewer branched rays in the anal fin; vomer toothed posteriorly but not anteriorly. 120–130 lateral cycloid scales. **Colour:** This is very variable, depending on the age, sex and condition of the fish. Young fish (fry and parr) are spotted with parr marks down each side. Smolts are very silvery. Adults vary from being very silvery with some dark spots when fresh-run, to multicoloured dark fish at spawning time. **Size:** 40–100 cm; maximum

120 cm; British rod record 29.029 kg (1922: River Tay). **Age:** Average adult age 4–9 years. **Distribution:** An anadromous native species widely found in the Atlantic areas of northern Europe and eastern North America. Widely distributed throughout the British Isles, and found in most unpolluted rivers and some large lakes (Map p. 197). **Habitat:** Clear stony rivers, streams and accessible lakes. **Reproduction:** October–January. Orange eggs (6 mm), laid in redds among gravel in running water, hatch after 70–160 days (depending on water temperature) into alevins, which change to fry, then to parr. After 2–6 years these become silver-coloured and migrate to the sea as smolts. Major growth occurs in the sea (fish returning within 1 year are called grilse; later than 1 year salmon). A 70 cm female carries 5,000–6,000 eggs. **Food:** Invertebrates when small, invertebrates and fish when larger. **Value:** A major commercial species, netted off Atlantic coasts and estuaries and off Greenland. An important sport species angled in rivers and lakes – fly-fishing, spinning and trolling are all successful in various waters. A major farmed species in several countries (e.g. Norway, Scotland and Ireland). **Conservation:** Because of its historic and economic importance, this species is given substantial protection in several pieces of British legislation. It is also listed in Annexes IIa and Va of the EC Habitat and Species Directive and in Appendix III of the Bern Convention. **Literature:** Aprahamian et al. (1998), Buck & Hay (1984), Buck & Youngson (1982), Calderwood (1930), Carss et al. (1990), Child et al. (1976), Clifford et al. (1998), Garcia & Verspoor (1989), Gardiner (1974), Gee et al. (1978), Hansen & Pethon (1985), Hindar & Nordland (1989), Hurnell & Price (1991), Jones (1959), L'Abbe-Lund (1988), MacCrimmon & Gots (1979), Malloch (1910), Mills (1964, 1971), Myers & Hutchings (1987), Payne et al. (1971), Perez et al. (1999), Pope et al. (1961), Pyefinch (1955), Solomon & Child (1978), Thorpe (1977a, 1987), Verspoor (1988), Youngson et al. (1992).

Brown Trout *Salmo trutta* (Figs 47, 51, 53; Plate 32)

Distinctive features: Vomer toothed anteriorly and posteriorly; 110–120 cycloid scales along the lateral line. Numerous subspecies have been described and there are many local names (e.g. Finnock, Whitling). **Colour:** This is very variable according to age, season and location. Typically, in Brown Trout the back is brownish grey, grading to silvery yellow on the sides and yellowish white on the belly. Superimposed on this is a wide variety of large and small spots, mostly black, but some red. Sea Trout, after returning from the sea, are very silvery with dark spots. **Size:** 15–50 cm; maximum 70 cm;

British rod record 13.863 kg (2000: Loch Awe). **Age:** Average adult age 8–15 years. **Distribution:** Occurs all over northern Europe. Migratory populations occur all along the European coast from Spain northwards. One of the commonest species in the British Isles, found in most unpolluted rivers and many lakes (Map p. 197). **Habitat:** Found in a wide variety of running and standing waters. **Reproduction:** Mainly October in small rivers and streams, where eggs are laid in a nest (redd) which the female cuts out in the gravel. The orange eggs (4 mm) hatch in approximately 150 days and the fry spend a year or more in the nursery stream before moving down into a larger river or lake (or the sea, in the case of the migratory Sea Trout). Adults mature after 3–5 years and many live for up to 20 years. The mean number of eggs is about 5,400 per female. **Food:** Invertebrates when young, invertebrates and some fish (especially in the case of Sea Trout) when adult. **Value:** An important commercial species in some areas, and possibly the most important single sport species over Europe as a whole. Widely introduced to many parts of the world, including North America, where it is well established in some areas. **Conservation:** This native species is one of the commonest fish in the British Isles and at present there are no national conservation measures. However, there are a number of valuable genetic stocks of this species which do require some measure of conservation. **Literature:** Allen (1938), Bagenal (1969a,b), Ball & Jones (1960), Bardonnet et al. (1993), Bembo et al. (1993), Burrough & Kennedy (1978), Calderwood (1930), Campbell (1977), Campbell (1957, 1963, 1971, 1979a), Crozier & Ferguson (1986), Egglishaw (1967), Eklov et al. (1999), Elliott (1976, 1994), Fahy (1977, 1978), Ferguson & Mason (1981), Frost (1939), Frost & Brown (1967), Gerrish (1939), Graham & Jones (1962), Hamilton et al. (1989), Hesthagen & Jonsson (1998), Hynd (1964), Kennedy & Strange (1978), Le Cren (1985), MacCrimmon & Marshall (1968), Malloch (1910), Milner et al. (1978), Nall (1930), Stuart (1953, 1957), Thorpe (1974), Treasurer (1976), Went (1979).

Arctic Charr *Salvelinus alpinus* (Figs 47, 48; Plate 33)

Distinctive features: Vomer toothless posteriorly; body uniformly coloured with pale spots. 190–220 cycloid scales along the lateral line. Many different morphs occur, sometimes in the same lake where there may be migratory, normal and dwarf populations living together. **Colour:** Young fish are a silvery monochrome with lateral parr marks. Adults are variable according to sex and season. The back is greyish green, grading to greyish brown on the sides and whitish grey on the belly. Superimposed on this are a variety of yellow, cream, pink, red or orange spots. Females are usually duller than males which, during the breeding season, usually have a bright red flash along the lower half of the body. **Size:** 15–40 cm; maximum 88 cm; British rod record 4.309 kg (1995: Loch Arkaig). **Age:** Average adult age 6–12 years. **Distribution:** A holarctic species found in many northern catchments in the Northern Hemisphere. In Arctic seas anadromous stocks are found, maturing in the sea and entering rivers to spawn. These fish are larger than the freshwater forms, some of which are dwarf races. In the British Isles there are probably at least 250 populations, mostly in Scotland, but populations also occur in Ireland, Wales and England (Map p. 198). **Habitat:** Mainly in clear, cool lakes and rivers, and around the coast in northern areas. **Reproduction:** October–March, spawning among gravel in both lakes and rivers. After hatching the young move into lakes or the sea, maturing in 3–6 years. 560–7,300 yellow eggs (3.5 mm) per female. **Food:** Mainly invertebrates, especially planktonic crustaceans. Large fish in some populations may be piscivorous (even cannibalistic). **Value:** Netted in Arctic areas during the spawning migration and in many lakes. Important subsistence fisheries occur among Inuit people in North America. A sporting species in some lakes, including some in Britain and Ireland. **Conservation:** This native species has undergone significant decline in several parts of the British Isles and is in need of conservation action. Unfortunately, it does not yet receive any national legislative protection. **Literature:** Adams et al. (1998), Andrews & Lear (1956), Barbour & Einarsson (1987), Bean et al. (1996), Campbell (1979a), Campbell (1982, 1984), Friend (1956), Frost (1977), Frost & Kipling (1980), Gardner et al. (1988), George & Winfield (2000), Hardie (1940), Hartley et al. (1992a,b), Kipling (1984), Maitland et al. (1984), Partington & Mills (1988), Walker et al. (1988), Went (1971), Winfield et al. (2002a).

Brook Charr *Salvelinus fontinalis* (Figs 47, 49; Plate 34)

Distinctive features: Vomer toothless; body with a marked vermiculate pattern, with dark and light areas interwoven. 160–230 cycloid scales along the lateral line. Known in North America as Brook Trout. **Colour:** The back is greenish grey with well-marked lighter vermiculations. These grade to a lighter colour on the sides and greenish white on the belly. Superimposed on the back and sides are spots coloured cream, yellow and red (sometimes with a blue halo). **Size:** 20–35 cm; maximum 50 cm; British rod record 3.713 kg

(1998: Fontburn Reservoir). **Age:** Average adult age 5–9 years. **Distribution:** Native to western North America, but introduced to many parts of Europe, including Britain, where it has become established in a number of waters (Map p. 198). Absent from Ireland. **Habitat:** Clear lakes, streams and rivers. **Reproduction:** October–March among gravel in running waters. The yellow eggs (3.5 mm) hatch in spring and the young spend about 2 years in nursery areas before moving into larger rivers or lakes. Adults mature at 2–3 years of age. 100–5,000 eggs per female. **Food:** Mainly invertebrates, but fish are eaten by large adults. **Value:** A major sport species in North America, and popular in some places in Europe. Stocked for angling in a few waters in Britain. **Conservation:** As an alien species which is not threatened in areas where it is indigenous, no conservation measures are necessary in Britain. **Literature:** Bourke et al. (1997), Bridges & Mullan (1958), Cunjac & Power (1986), Danzmann et al. (1991), Gaudreault et al. (1986), Lever (1977), MacCrimmon & Campbell (1969), Reimers (1979), Robinson et al. (1976), Scott & Crossman (1973), Wurtsbaugh et al. (1975).

GRAYLING FAMILY: THYMALLIDAE

Grayling *Thymallus thymallus* (Fig. 54; Plate 28)

Distinctive features: Well-developed adipose fin and large dorsal fins. The spot pattern is specific to individual fish. 80–88 lateral cycloid scales. **Colour:** The back is dull grey, grading to greyish green on the sides and silvery white on the belly. There are a number of clearly defined black spots, mostly anteriorly, with a pattern individual to each fish. **Size:** 25–35 cm; maximum 50 cm; maximum weight 4.675 kg; British rod record 1.899 kg (1989: River Frome). **Age:** Average adult age 7–10 years. **Distribution:** Occurs over much of northern and central Europe. Native to England, but introduced elsewhere in England and Wales and now common in many rivers in Scotland from the River Tay southwards (Map p. 199). Absent from Ireland. **Habitat:** Found in clean, cool rivers (occasionally lakes), sometimes in estuaries. **Reproduction:** March–May, yellow eggs (3.6 mm) laid in nests in gravel hatch in 20–30 days and the young mature at 3–4 years. They live for up to 15 years. Females about 45 cm long carry about 10,000 eggs. **Food:** Mainly invertebrates, especially insects. Larger individuals may eat small fish. **Value:** of some commercial importance to fisheries in Europe, especially in Russia. A popular sport fish in some countries, including Britain, where it is caught on worms or artificial flies. **Conservation:** This native species is relatively common and at present there are no national conservation measures. However, it is given some protection by its listing in Annex Va of the EC Habitat and Species Directive and in Appendix III of the Bern Convention. **Literature:** Bardonnet & Gaudin (1990), Darchambeau & Poncia (1997), Fabricius & Gustavson (1955), Gerrish (1939), Hellawell (1969, 1971a), Hutton (1923), Jones (1953), Mackay (1970), Parkinson et al. (1999), Radforth (1940), Scott (1985), Woolland (1987), Woolland & Jones (1975).

COD FAMILY: GADIDAE

Burbot *Lota lota* (Fig. 55; Plate 35)

Distinctive features: Elongate body with two dorsal fins, the first shorter than the second; broad head with one long barbel on the lower jaw, and one shorter barbel at each nostril. Scales small, cycloid and embedded. **Colour:** The back is yellow-olive to olive-green with dark marbling, the colours and patterning being paler on the sides. The belly is yellowish white. **Size:** 30–50 cm; maximum 120 cm; maximum weight about 32 kg. **Age:** Average adult age 10–15 years. **Distribution:** Occurs throughout northern areas of Europe, Asia and North America. Has declined in some parts of its range including Britain and it is now extinct in those catchments in southeast England where it formerly occurred (Map p. 199). Absent from Ireland. **Habitat:** Found in cool, clear rivers and lakes. **Reproduction:** December–March, over stones and gravel in rivers and lakes. Adhesive pale yellow eggs (1.1 mm) hatch in 40–50 days and the young mature after 3–4 years. Can live for up to 25 years. 33,100–3,063,000 eggs per female. **Food:** When young, invertebrates (especially crustaceans and insect larvae); when older, fish. **Value:** Of some commercial value in Europe and North America, caught by means of nets, traps and baited hooks. Of some sporting value for angling through ice. **Conservation:** Because of its decline in parts of Europe, the Burbot is now given some protection. It is listed in Schedule 5 of the 1981 Wildlife and Countryside Act (rather late since it appears to have become extinct by the 1970s) and in Annexes IIa and Va of the EU Habitats and Species Directive, and as a Long List Species in the UK Biodiversity Action Plan. There is no Red Data Book for fish in Britain, and it is not listed in the EC Habitat and Species Directive, but Maitland (2000) considers this species to be

Extinct in Great Britain. **Literature:** Bailey (1972), Clemens (1951a,b), Ghan & Sprules (1993), Hinkens & Cochrane (1988), Lawler (1963), MacCrimmon & Devitt (1954), Marlborough (1970), Palliainen & Korhonen (1990, 1993).

GREY MULLET FAMILY: MUGILIDAE

Thick-lipped Grey Mullet *Chelon labrosus* (Fig. 56; Plate 45)

Distinctive features: Body covered with large cycloid scales which extend on to the head; thick adipose eyelid; two dorsal fins, the first with only 4 spiny rays; upper lip very thick and with two rows of small warts; 45–46 lateral scales. **Colour:** The back is dark greenish grey-blue which grades to silver with longitudinal grey stripes along the sides, and pure white on the belly. **Size:** 30–50 cm; maximum 90 cm; British rod record 6.428 kg (1979: Aberthaw, Glamorgan). **Age:** Average adult age 8–10 years. **Distribution:** Found in Atlantic coastal areas south of Norway and in the Mediterranean and Black Seas. Occurs around much of the coasts of Britain and Ireland, especially in southern areas (Map p. 200). **Habitat:** Occurs in fresh water in the lower reaches of some rivers. **Reproduction:** June–August, in the sea, where the clear eggs (1.0 mm) float until they hatch. The young mature after 2–4 years. **Food:** Filamentous algae, diatoms and other bottom-living plants, which they eat by ingesting the surface layer of soft sediments. **Value:** Of considerable commercial value (along with other mullets) in net and trap fisheries around Europe. Considered to be a sport species in some areas. **Conservation:** This native species is relatively common and there are no national conservation measures at present. **Literature:** Almeida et al. (1993), Anderson (1982), Biagianti-Risbourg (1991), De Silva (1980), Erman (1961), Farrugio (1977), Flowerdew & Grove (1980), Guinea & Fernandez (1992), Hickling (1970), Kennedy & Fitzmaurice (1969), Reay & Cornell (1988), Romer & McLachlan (1986).

Golden Grey Mullet *Liza aurata* (Fig. 57; Plate 46)

Distinctive features: Body covered with large cycloid scales which extend on to the head; operculum and cheek with golden patches; two dorsal fins, the first with only 4 spiny rays; no elongate lobule above the base of the pectoral fin; golden spot on the gill cover and behind each eye. **Colour:** The back is dark greyish blue which grades to silvery gold with longitudinal greyish gold stripes along the sides, and pure white on the belly. **Size:** 20–35 cm; maximum 50 cm; British rod record 1.368 kg (1994: Christchurch, Dorset and another off Alderney, 1991). **Age:** Average adult age 6–8 years. **Distribution:** Found all round the Atlantic coastal areas of southern Europe, including the Mediterranean and Black Seas. It has been successfully introduced to the Caspian Sea. Occurs around much of the coasts of Ireland and Britain, especially in southern areas (Map p. 200). **Habitat:** Coastal areas and the lower reaches of large rivers. **Reproduction:** August–September, spawning in the sea where the clear eggs (1.0 mm) float until they hatch. The young mature after 3–5 years and may live for up to 10 years. 1,200,000–2,100,000 eggs per female. **Food:** Filamentous algae, diatoms and other bottom-living plants, which they eat by ingesting the surface layer of soft sediments. **Value:** Of considerable commercial value in net and trap fisheries, especially in the Mediterranean and Black Seas. Of little sporting significance. **Conservation:** This native species is relatively common and there are no national conservation measures at present. **Literature:** Anderson (1982), Biagianti-Risbourg (1991), De Silva (1980), Farrugio (1977), Guinea & Fernandez (1992), Hickling (1970), Reay & Cornell (1988).

Thin-lipped Grey Mullet *Liza ramada* (Fig. 58)

Distinctive features: Upper lip with only one row of small tubercles; body covered with large cycloid scales which extend on to the head; two dorsal fins, the first with only 4 spiny rays; no golden spots on the gill covers. **Colour:** The back is dark greenish grey-blue which grades to silver with longitudinal grey stripes along the sides, and pure white on the belly. **Size:** 25–40 cm; maximum 60 cm; British rod record 3.175 kg (1991: Oulton Broad). **Age:** Average adult age 8–10 years. **Distribution:** Found round all parts of the European coast (including the Mediterranean and Black Seas) except the extreme north. Occurs around much of the coasts of Ireland and Britain, especially in southern areas (Map p. 201). **Habitat:** The sea and coastal waters, entering the lower reaches of large rivers and some coastal lakes and lagoons. **Reproduction:** August–September, spawning in the sea where the clear eggs float until they hatch. The young mature after 3–5 years. **Food:** Filamentous algae, diatoms and other bottom-living plants, which they eat by ingesting the surface layer of soft sediments. **Value:** Of considerable commercial and some sporting value in various parts of its range. **Conservation:** This native species is relatively common and there are no national conservation measures at present. **Literature:** Anderson (1982), Biagianti-Risbourg (1991), De Silva (1980), Farrugio (1977), Guinea & Fernandez (1992), Hickling (1970), Reay & Cornell (1988).

STICKLEBACK FAMILY: GASTEROSTEIDAE

Three-spined Stickleback *Gasterosteus aculeatus* (Fig. 59; Plate 37)

Distinctive features: Three strong spines anterior to the dorsal fin; no scales on the body but this may be protected by a variable number of bony plates. In some parts of northwest Scotland, spineless forms occur. **Colour:** In the sea, this species has a light olive to grey-green back and bright silver flanks. Freshwater forms, however, tend to have olive to brownish backs with grey-brown mottling on the silvery flanks. At spawning time the male develops a bright red throat and is iridescent green elsewhere. The eye is a brilliant aquamarine. **Size:** 4–8 cm; maximum 11 cm. British rod record 7 g (1998: High Flyer Lake). **Age:** Average adult age 2–3 years. **Distribution:** Found in many parts of Europe, especially areas which are not too far from the sea. Indigenous to the British Isles (Map p. 201). **Habitat:** Occurs in a wide variety of waters from the sea to estuaries, rivers and lakes of all kinds. **Reproduction:** March–June, when the male builds a nest of fibrous material and induces one or more females to lay in it. The yellow eggs (1.5 mm) hatch in 5–20 days and they, and the young, are guarded by the male. The young mature after 1–2 years and rarely live beyond 4 years. 90–450 eggs per female. **Food:** Invertebrates (mainly worms, crustaceans and insect larvae) and sometimes small fish. **Value:** Formerly used in parts of Europe for the production of fish meal, it is now of little commercial or sporting significance. It is commonly kept in aquaria, however, and is often used for teaching purposes. **Conservation:** This native species is relatively common and there are no national conservation measures at present. **Literature:** Allen & Wootton (1982), Campbell (1979b, 1985), Campbell (1984), Chappell (1969), Giles (1983), Greenbank & Nelson (1959), Hynes (1950), Ibrahim & Huntingford (1988), Jones & Hynes (1950), Krause (1993), Lewis et al. (1972), Whoriskey et al. (1986), Wootton (1976, 1984), Wright & Huntingford (1993).

Nine-spined Stickleback *Pungitius pungitius* (Fig. 60; Plate 38)

Distinctive features: 7–12 (although usually 9) stiff spines anterior to the dorsal fin; body without scales. At spawning time the male develops a black throat and becomes very dark elsewhere. Also known as Ten-spined Stickleback. **Colour:** The back and flanks are usually a greenish olive-gold, grading to a pale white belly. At spawning time the male develops a sooty black head and throat, this colour sometimes extending to much of the body. **Size:** 5–7 cm; maximum 9 cm. **Age:** Average adult age 2–3 years. **Distribution:** Found in most parts of northern Europe, Asia and North American which are not too far from the sea. Its distribution in the British Isles is not well documented, but is sporadic and it is certainly rare in most parts of Scotland (Map p. 202). **Habitat:** Occurs in both brackish and fresh waters, including both slow-flowing streams and lakes, especially where there is a lot of weed cover. **Reproduction:** April–July; the male builds a nest of fine plant material among vegetation, and induces one or more females to lay in it. The yellow eggs (1.2 mm) hatch after 10–20 days; both they and the fry are guarded by the male. The young mature after 1 year and live for only 2–3 years at most. **Food:** Invertebrates, especially crustaceans and insect larvae. **Value:** Of no commercial or sporting value. Sometimes kept in aquaria. **Conservation:** Regarded by many as a common species, the distribution of this species is not well known and it has certainly declined in some areas. Local action is needed in Scotland and some other parts of Britain. **Literature:** Campbell (1979b), Dartnell (1973), Griswold & Smith (1973), Hynes (1950), Jones & Hynes (1950), Lewis et al. (1972), McKenzie & Keenleyside (1970), Morris (1952), Solanki & Benjamin (1982), Wootton (1976, 1984).

SCULPIN FAMILY: COTTIDAE

Common Bullhead *Cottus gobio* (Fig. 61; Plate 47)

Distinctive features: No scales; lateral line with 30–35 pores and ending at the caudal fin. Known also as Miller's Thumb. **Colour:** The back is pale brownish grey, paling down the sides to muddy white on the belly. However, the whole body is covered with a dark mottling, the intensity of which varies with the background on which the fish is resting and so is very variable. **Size:** 10–15 cm; maximum 18 cm. British rod record 28.4 g (1983: Green River).

Age: Average adult age 3–5 years. **Distribution:** Found across most of Europe except the extreme north and south. Indigenous to England and possibly Wales, but introduced to a few catchments in Scotland. Absent from Ireland (Map p. 202). **Habitat:** Occurs mainly in stony streams but also in some oligotrophic lakes. **Reproduction:** March–May, spawning in nests under stones guarded by the male. The pale yellow eggs (2.3 mm), laid in clumps, hatch in 20–25 days and young mature in 2 years, rarely living longer than 6 years. About 100 eggs per female. **Food:** Mainly invertebrates, especially insect larvae, but also fish eggs and fry. **Value:** Of no commercial or sporting value. **Conservation:** This native species is relatively common and there are no national conservation measures at present. However, because it has declined in parts of Europe, it is given some protection by its listing in Annex IIa of the EC Habitat and Species Directive. **Literature:** Clelland (1971), Crisp (1963), Crisp et al. (1975), Fox (1978), Hyslop (1982), Ladich (1989), Marconato & Bisazza (1988), Morris (1978), Rumpus (1975), Smyly (1957), Western (1971).

BASS FAMILY: MORONIDAE

Sea Bass *Dicentrarchus labrax* (Fig. 62; Plate 39)

Distinctive features: Two equal dorsal fins, which just meet each other; scales between the eyes are cycloid; 52–74 ctenoid scales along the lateral line; large dark mark on the gill cover, but none on the body. **Colour:** The back is greyish green to bluish, shading to bright silver along the sides and silvery white along the belly. **Size:** 40–70 cm; maximum 100 cm; maximum weight about 12 kg; British rod record 8.877 kg (1987: Herne Bay). **Age:** Average adult age 15–25 years. **Distribution:** Found all round Europe (including the Mediterranean and Black Seas) except in the extreme north. Occurs around much of the coasts of Britain and Ireland, especially in southern areas (Map p. 203). **Habitat:** Coastal areas and the lower reaches of large rivers. **Reproduction:** April–July, in the sea offshore, laying clear planktonic eggs (1.0 mm). Large females may have up to 2,000,000 eggs. Mature at 3–4 years and may live for up to 30 years. **Food:** When young, invertebrates (especially molluscs and crustaceans) and some fish; when adult mainly fish (e.g. herring). **Value:** Of minor commercial importance in some countries. An important sporting species in Britain and Ireland, offering exciting angling in some estuaries and coastal waters. **Conservation:** This native species is relatively common and at present there are no national conservation measures other than those related to commercial fishing at sea. **Literature:** Castilho & McAndrew (1998), Dando & Demir (1985), Holden & Williams (1974), Jackman (1954), Kelley (1979, 1986), Kennedy & Fitzmaurice (1968b, 1972b), Mayer et al. (1988, 1990), Pawson & Eaton (1999), Pawson & Pickett (1987), Pawson et al. (1987), Thompson & Harrop (1987).

SUNFISH FAMILY: CENTRARCHIDAE

Rock Bass *Ambloplites rupestris* (Fig. 64; Plate 40)

Distinctive features: Anal fin with 6 spines arising in a scaled groove; 7–9 horizontal rows of black spots below the lateral line; 39–40 lateral ctenoid scales. **Colour:** Greenish olive to golden brown on the back and sides with the ventral surface silvery white. Some dark blotchings on the upper sides which vary greatly in intensity. **Size:** 15–20 cm; maximum 34 cm; maximum weight 1.7 kg. **Age:** Average adult age 5–9 years. **Distribution:** Native to eastern central North America, it has been introduced to Europe and has become established in southern England, where there is one long-established population (Map p. 203). **Habitat:** Weedy lakes and the lower reaches of rivers. **Reproduction:** May–July, spawning in a nest excavated by the male among sand and gravel. The male guards the eggs (which hatch in 3–4 days) and early fry. Young fish mature in 2–3 years and live for up to 10 years. 3,000–11,000 yellow eggs (2.0 mm) per female. **Food:** Invertebrates (especially crustaceans and insect larvae) and small fish. **Value:** Of no significance in Europe but a valuable commercial and sporting fish in North America. **Conservation:** As an alien species which is not threatened in areas where it is indigenous, no conservation measures are necessary in Britain. **Literature:** Carlander (1977), George & Hadley (1979), Hile (1941), Jenkins & Burkhead (1993), Noltie & Keenleyside (1986), Petrimoulx (1984), Probst et al. (1984), Scott (1949), Scott & Crossman (1973), Vadas (1990).

Pumpkinseed *Lepomis gibbosus* (Fig. 65; Plate 41)

Distinctive features: Single dorsal fin divided into two parts; gill rakers knobbed; opercular flap short with a prominent red spot posteriorly; 40–47 lateral ctenoid scales. **Colour:** Very variable in colour according to age, sex and season. The back is greenish bronze merging to dull gold or greenish blue on the sides and yellowish white on the belly. The back and sides are overlain by a dark brownish black freckling and there may be an overall iridescence. The colours intensify at spawning time. **Size:** 10–15 cm; maximum 22 cm; maximum weight about 300 g. British rod record 129.4 g (1987: Whessoe Pond) **Age:** Average adult age 4–8 years. **Distribution:** Native to fresh waters in eastern North America, from Canada to Tennessee, this species has been introduced and become established in many parts of Europe, including southern England where there are several populations (Map p. 204). **Habitat:** Weedy lakes and the lower reaches of rivers. **Reproduction:** May–July, spawning in shallow depressions among sand in weed beds. The yellow eggs (2.0 mm), which hatch in 3–5 days, and the fry, are guarded by the male. The young mature after 2–3 years and may live for 9 years. 600–5,000 eggs per female. **Food:** Mainly invertebrates (especially crustaceans and insect larvae) but small fish are eaten by adults. **Value:** Of little significance in Europe, but caught in some numbers in North America, both commercially and as a sport fish. **Conservation:** As an alien species, which is not threatened in areas where it is indigenous, no conservation measures are necessary in Britain. **Literature:** Carlander (1977), Etnier (1971), Graham & Hastings (1984), Hardy (1978), Keast & Welsh (1968), Lever (1977), O'Hara (1968), Reid (1930), Scott & Crossman (1973), Tin (1982b), Wheeler & Maitland (1973).

Largemouth Bass *Micropterus salmoides* (Fig. 63; Plate 42)

Distinctive features: Upper jaw extending behind the eye; 60–68 lateral ctenoid scales. **Colour:** The back is dark olive, shading to pale olive-green on the sides and whitish on the belly. A wide irregular band of single dark blotches runs along the flanks from head to tail. **Size:** 20–40 cm; maximum 83 cm; maximum weight 11 kg; world rod record 10.092 kg. **Age:** Average adult age 8–12 years. **Distribution:** Native to southern Canada and USA, it has been introduced to Europe and is established in a number of countries, including England. However, there have been no recent records from populations previously reported in Surrey and Dorset (Map p. 204). **Habitat:** Weedy lakes and the lower reaches of rivers. **Reproduction:** March–July, spawning in pits dug out in sand and gravel. The yellow eggs (2.0 mm) are guarded by the male and take 2–5 days to hatch. Young mature after 3–4 years and live for up to 15 years. 750–11,500 eggs per female. **Food:** Invertebrates (especially crustaceans and insect larvae) when young, large invertebrates, fish and frogs when adult. **Value:** Of relatively little – but increasing – value in Europe, but a prized sport fish in North America where it is also caught commercially. **Conservation:** As an alien species which is not threatened in areas where it is indigenous, no conservation measures are necessary in Britain. **Literature:** Diana (1984), Galloway & Kilambi (1988), Hazen & Esch (1978), Keast & Eadie (1985), Lewis (1965), Maitland & Price (1969), Mraz et al. (1961), Philipp et al. (1985), Scott & Crossman (1973), Smagula & Adelman (1983), Spoor (1977).

PERCH FAMILY: PERCIDAE

Ruffe *Gymnocephalus cernuus* (Fig. 66; Plate 43)

Distinctive features: Two dorsal fins partially joined to one another, the anterior with 11–16 spines; 35–40 lateral ctenoid scales. Formerly *Gymnocephalus cernua*. **Colour:** The back and sides are sandy to pale brownish green with irregular dark blotches. The belly is pale yellow. **Size:** 10–15 cm; maximum 30 cm (exceptionally 50 cm in Siberia); British rod record 142 g (1980: West View Farm). **Age:** Average adult age 3–5 years. **Distribution:** Found throughout much of Europe, except certain areas in the north and south. Indigenous to southeast England but introduced to Wales and southern Scotland. Absent from Ireland (Map p. 205). Introduced to – and expanding rapidly in – Lake Superior, through the discharge of ballast water from ships entering the Great Lakes from Europe. **Habitat:** Occurs in slow-flowing rivers, canals and lakes. **Reproduction:** April–May, spawning among stones and vegetation in shallow water. The whitish yellow eggs (0.8 mm) hatch in 8–12 days and the young mature after 2–3 years. 4,000–104,000 eggs per female. **Food:** Bottom-dwelling invertebrates (especially molluscs, crustaceans and insect larvae) and sometimes fish eggs and small fish. **Value:** Of minor commercial and sporting value, although it is caught in a few areas and used as live bait in others. **Conservation:** This native species is relatively common and there are no national conservation measures at present. **Literature:** Adams & Maitland (1998), Adams & Tippett (1991), Fedorova & Vetkasov (1973), Hartley (1947a,b, 1948), Kolomin (1977), Maitland & East (1989), Maitland et al. (1983), Oliva & Vostradovsky (1960), Simon & Vondruska (1991), Wheeler (1969), Winfield et al. (1996, 1998b,c).

European Perch *Perca fluviatilis* (Fig. 67; Front cover illustration)

Distinctive features: Separate dorsal fins, the first very spiny; reddish pectoral and pelvic fins; several dark vertical stripes on sides. 58–68 lateral ctenoid scales. **Colour:** The back is grey to olive-green, grading to pale olive on the flanks and yellowish white on the belly. Several well-marked dark bands run vertically from the back down each side. **Size:** 20–35 cm; maximum 51 cm; maximum weight about 4.75 kg; British rod record 2.523 kg (1985: undisclosed lake). **Age:** Average adult age 8–12 years. **Distribution:** Occurs over most of northern Europe and Asia. Indigenous to much of the British Isles, though absent from the extreme northwest and most islands (Map p. 205). The Yellow Perch *Perca flavescens* of eastern North America is believed by some to be the same species. **Habitat:** Found in slow-flowing rivers, ponds and lakes. **Reproduction:** April–June, in shallow water among vegetation. The whitish eggs (2.3 mm), laid in ribbons about 1 m long, hatch in 15–20 days. Young mature in 2–3 years and live for up to 10 years. 12,000–199,000 eggs per female. **Food:** Invertebrates (especially crustaceans and insect larvae) when young, invertebrates and fish when older. **Value:** Caught commercially in some countries by traps, nets and baited lines. Very important as a sport species in Britain and Ireland, and other parts of Europe. **Conservation:** This native species is relatively common and there are no national conservation measures at present. **Literature:** Andrews (1979), Bodaly et al. (1989), Bregazzi & Kennedy (1982), Burrough & Kennedy (1978), Campbell (1955), Coles (1981), Craig (1977, 1987), Craig & Kipling (1983), Goldspink (1990), Goldspink & Goodwin (1979), Jones (1953), Karas (1990), Lang (1987), Le Cren et al. (1967), Rask et al. (1990), Shafi (1969), Shafi & Maitland (1971b), Thorpe (1977b), Treasurer (1981, 1989), Williams (1965), Willoughby (1970).

Pikeperch *Sander lucioperca* (Fig. 68; Plate 44)

Distinctive features: Dorsal fins almost touching; large canine teeth; 80–95 lateral ctenoid scales. Also called Zander and formerly *Stizostedion lucioperca*. **Colour:** The back and flanks are greenish brown or greyish with a gold sheen. The belly is white. Several irregular dark bands run vertically from the back down each side. **Size:** 30–70 cm; maximum 130 cm; maximum weight about 18 kg; British rod record 8.774 kg (1998: Fen Drain). **Age:** Average adult age 10–15 years. **Distribution:** Found across much of Europe from the Netherlands to the Caspian Sea. Introduced to other areas, including England, where it is slowly spreading from its original strongholds to other catchments, due to the activities of anglers (Map p. 206). **Habitat:** Occurs in slow-flowing weedy rivers, canals and rich lakes **Reproduction:** April–June, among gravel and stones. The whitish yellow eggs (1.3 mm) are guarded by both parents and hatch in 5–10 days. Young mature in 3–5 years. 180,000–1,185,000 eggs per female. **Food:** Invertebrates at first, but almost entirely fish thereafter. **Value:** Taken commercially in traps and nets in Europe and a valued food fish there. A prized sport species in England and some other countries in Europe. **Conservation:** As an alien species which is not threatened in areas where it is indigenous, no conservation measures are necessary in Britain. **Literature:** Fickling & Lee (1985), Linfield & Rickards (1979), Mansfield (1958), Puke (1952), Rickards & Fickling (1979), Steffens (1960), Svardson & Molin (1973), Ziukov & Petrova (1993).

GOBY FAMILY: GOBIIDAE

Common Goby *Pomatoschistus microps* (Fig. 69; Plate 48)

Distinctive features: Two dorsal fins and united pelvic fins; 42–52 lateral ctenoid scales; space between the dorsal fins is much less than the length of the first dorsal fin. **Colour:** The back is light brownish grey, grading to cream underneath. Overlying the upper part of the body there is a complex network of dark spots, with a single row of dark marks along each side. **Size:** 3–6 cm; maximum 7 cm. **Age:** Average adult age 1–2 years. **Distribution:** Occurs round the coasts of Europe from southern Norway to the Mediterranean and Black Seas. It is found all round the shores and estuaries of the British Isles from Shetland to Cornwall. **Habitat:** Found in shallow coastal areas, and often in brackish-freshwater estuaries (Map p. 206). Of all the British gobies, the Common Goby is the most likely species to penetrate fresh water. **Reproduction:** April–September, spawning in a nest cleared by the male under a shell or stone. The male guards the clear oval eggs (1.0 mm) and fans water over them until they hatch; fish may spawn up to 8 times in a breeding season. The young mature after 1 year, and few fish live beyond 2 years. **Food:** Mainly benthic invertebrates, especially crustaceans. **Value:** Of no commercial or sporting value. **Conservation:** Though listed in Appendix III of the Bern Convention, this native species is relatively common and there are no national conservation measures at present. **Literature:** Al Hassan et al. (1987), Fouda (1979), Fouda & Miller (1979), Healey (1972), Magnhagen (1998), Miller (1975), Rogers (1988).

FLATFISH FAMILY: PLEURONECTIDAE

Flounder *Platichthys flesus* (Fig. 70)

Distinctive features: Flattened asymmetrically with both eyes on the same side – usually the right, which is uppermost when lying on the bottom, but occasionally specimens are found with eyes on the left side; bony tubercles and platelets on the body and a well-developed lateral line; 15–22 gill rakers on the first arch. Scales are irregular and rough to touch – the bases of dorsal and ventral fins are also rough – a simple tactile method of distinguishing Flounder from other flatfish. Some workers now place this species in the genus *Pleuronectes*. **Colour:** Upperside dull brown with indistinct grey-brown blotches and dull red spots; underside pale greyish white. **Size:** 20–30 cm; maximum 50 cm; British rod record 2.594 kg (1956: Fowey, Cornwall). **Age:** Average adult age 5–10 years. **Distribution:** Found all round the European coast from the Arctic Ocean to the northern Mediterranean and Black Seas. Occurs around the entire coastlines of Britain and Ireland (Map p. 207). **Habitat:** As well as the sea, this species is common also in estuaries and lowland rivers, and in some lakes that are easily accessible from the sea. **Reproduction:** February–May, spawning in the sea in deep water. The planktonic clear eggs (0.6 mm) hatch in 4–8 days and the larvae are pelagic for about 50 days before sinking to the bottom and developing their flattened form. The young mature after 3–4 years. 500,000–2,000,000 eggs per female. Hybridises with Plaice, *Pleuronectes platessa* (Linnaeus 1758). **Food:** Zooplankton when young, benthic invertebrates, especially worms, molluscs and crustaceans when older. **Value:** Of considerable commercial value in many sandy coastal areas around the British Isles, where it is caught in traps and nets of various types. It is also popular with anglers in many coastal areas. **Conservation:** This native species is relatively common and there are no national conservation measures at present other than those related to commercial fishing in the sea. **Literature:** Beaumont & Mann (1983), Carter et al. (1998), Gibson (1972), Hutchinson & Hawkins (1993), Jones (1952), Kennedy (1984a), Lorenzen et al. (1991), Moore & Moore (1976), Mulicki (1947), Munro et al. (1989), Radforth (1940), Reiersen & Fugelie (1984), Summers (1979), Van Den Broek (1979).

DISTRIBUTION MAPS OF FISH OCCURRING IN THE FRESH WATERS OF BRITAIN AND IRELAND

HYDROMETRIC AREAS

The distribution maps given on pages 180 to 207 show the general occurrence of each species in Britain and Ireland, based on records for Hydrometric Areas (pages 178-179). A Hydrometric Area is a river catchment having one or more outlets to the sea or tidal estuary. For convenience some Hydrometric Areas include several smaller catchments with separate tidal outlets where there is topographical similarity (Water Resources Board & Scottish Development Department 1974; Smith & Lyle 1979; Environmental Protection Agency 2001). Some Areas are relatively large (e.g. Tay, Trent and Yorkshire Ouse) whereas others are small (e.g. Isle of Wight, Anglesey and Isle of Man).

More detailed maps of fish distribution in Britain (not Ireland), of the type included in the first edition of this key (Maitland 1969a, 1972a), are available in Davies et al. (2004). The distribution of many species is changing – new information would be welcome and should be passed to appropriate local museums or biological record centres. Several species are under-recorded and this is indicated on maps where the species is believed to be more widespread than has been shown here.

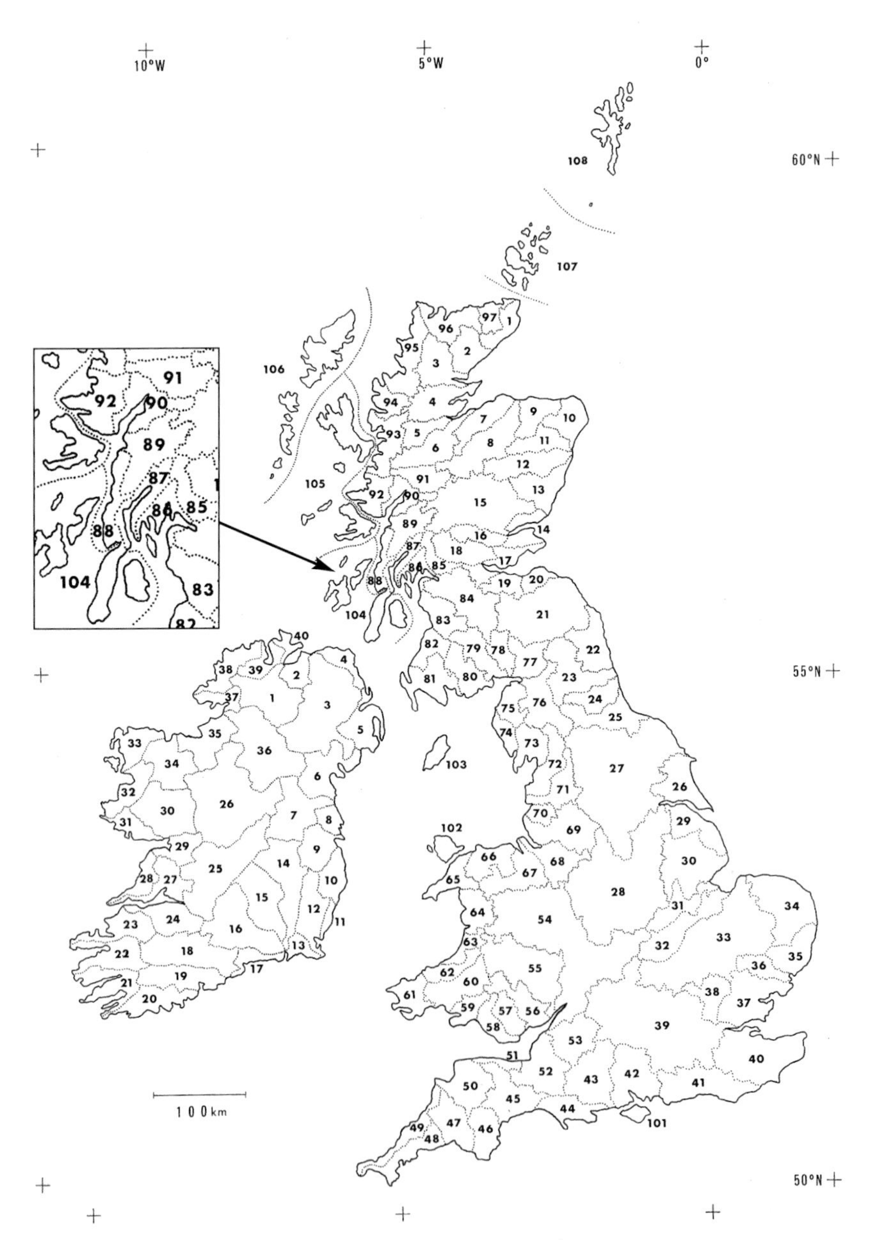
10°W
5°W
0°
60°N
55°N
50°N
100 km

Hydrometric Areas in Britain

1	Wicks Group	2	Helmsdale Group	3	Shin Group
4	Conon Group	5	Beauly	6	Ness
7	Findhorn Group	8	Spey	9	Deveron Group
10	Ythan Group	11	Don (Grampian)	12	Dee (Grampian)
13	Esk Group	14	Firth of Tay Group	15	Tay
16	Earn	17	Firth of Forth Group	18	Forth
19	Almond Group	20	Tyne (Lothian) Group	21	Tweed
22	Coquet Group	23	Tyne (Northumberland)	24	Wear
25	Tees Group	26	Hull Group	27	Ouse (Yorkshire)
28	Trent	29	Ancholme Group	30	Witham & Steeping
31	Welland	32	Nene	33	Great Ouse
34	Norfolk Rivers Group	35	East Suffolk Rivers	36	Stour (Essex & Suffolk)
37	Essex Rivers Group	38	Lee	39	Thames
40	Kent Rivers Group	41	Sussex Rivers Group	42	Hampshire Rivers Group
43	Avon and Stour	44	Frome Group	45	Exe Group
46	Dart Group	47	Tamar Group	48	Fal Group
49	Camel Group	50	Taw & Torridge	51	East Lyn Group
52	Somerset Rivers Group	53	Avon (Bristol)	54	Severn
55	Wye (Hereford)	56	Usk	57	Taff (Glamorgan) Group
58	Mid Glamorgan Group	59	Loughor Group	60	Towy
61	Cleddau Group	62	Teifi	63	Ystwyth Group
64	Dyfi Group	65	Glasslyn Group	66	Conway & Clwyd
67	Dee (Cheshire)	68	Cheshire Rivers Group	69	Mersey & Irwell
70	Douglas Group	71	Ribble	72	Wyre & Lune
73	Kent Group	74	Esk (Cumbria) Group	75	Derwent (Cumbria) Group
76	Eden (Cumbria)	77	Esk (Dumfries)	78	Annan
79	Nith	80	Dee (Galloway)	81	Cree Group
82	Doon Group	83	Irvine & Ayr	84	Clyde
85	Leven (Strathclyde)	86	Firth of Clyde Group	87	Fyne Group
88.	Add Group (Knapdale)	89	Awe and Etive	90	Loch Linnhe Group
91	Lochy (Highlands)	92	Loch Shiel Group	93	Loch Alsh Group
94	Loch Maree Group	95	Laxford Group	96	Naver Group
97	Thurso Group				
101	Isle of Wight	102	Anglesey	103	Isle of Man
104	Kintyre Group	105	Inner Hebrides	106	Outer Hebrides
107	Orkneys	108	Shetlands		

Hydrometric Areas in Ireland

1	Foyle	2	Faughan-Roe	3	Bann
4	Bush & N.E. Streams	5	Lagan-Quoile	6	Newry-Fane-Glyde-Dee
7	Boyne	8	Nanny-Delvin	9	Liffey-Dublin Bay
10	Avoca-Vartry	11	Owenavorragh	12	Slaney-Wexford Harbour
13	Ballyteigue-Bannow	14	Barrow	15	Nore
16	Suir	17	Colligan-Mahon	18	Blackwater (Munster)
19	Lee, Cork Harbour and Youghal Bay			20	Bandon-Ilen
21	Dunmanus-Bantry-Kenmare			22	Laune-Maine-Dingle Bay
23	Tralee Bay-Feale	24	Shannon Estuary South	25	Lower Shannon
26	Upper Shannon	27	Shannon Estuary North	28	Mal Bay
29	Galway Bay South East	30	Corrib	31	Galway Bay North
32	Erriff-Clew Bay	33	Blacksod-Broadhaven	34	Moy-Killala Bay
35	Sligo Bay-Drowes	36	Erne	37	Donegal Bay North
38	Gweebarra-Sheephaven	39	Lough Swilly	40	Donagh-Moville

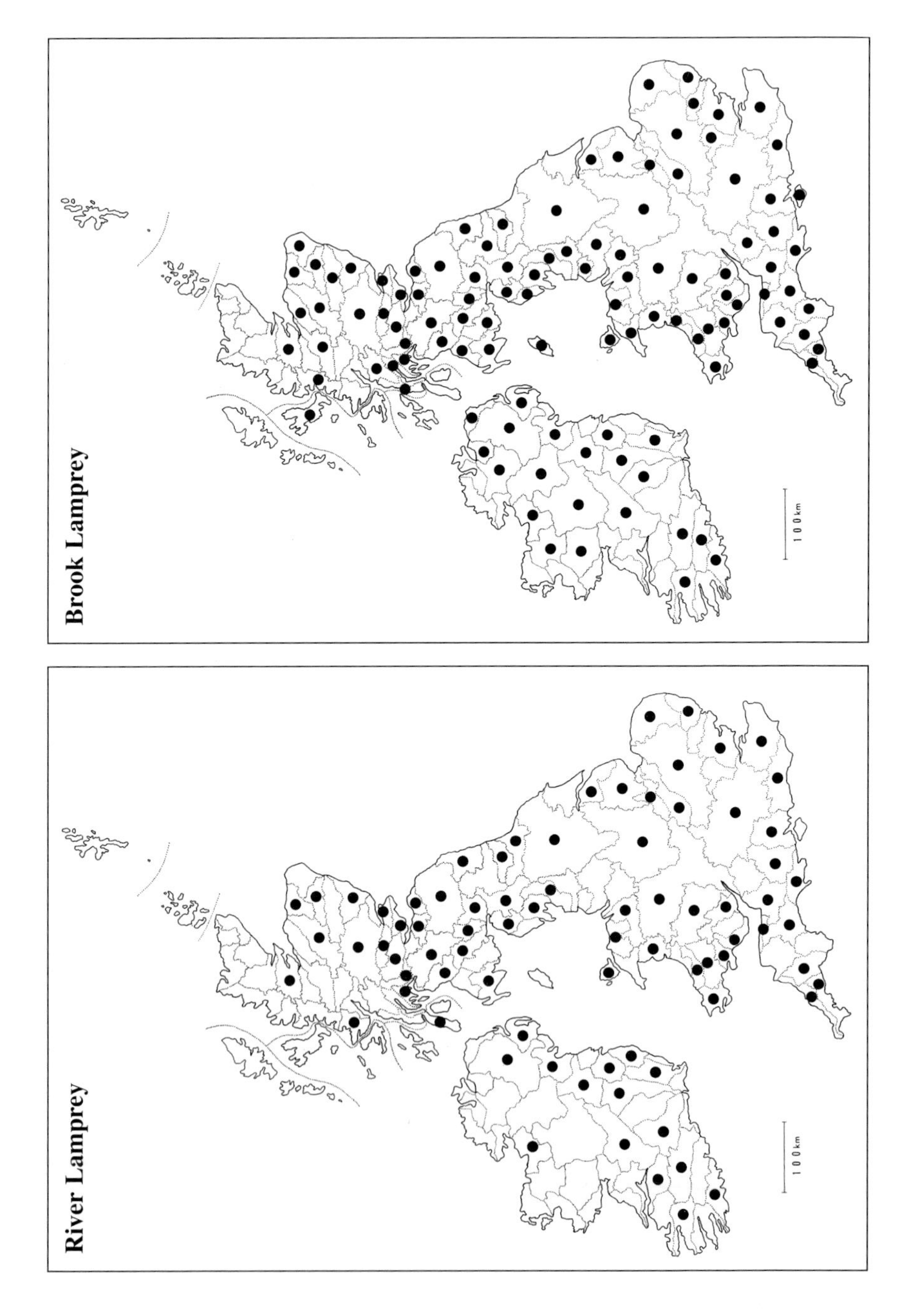
Brook Lamprey
100 km
River Lamprey
100 km

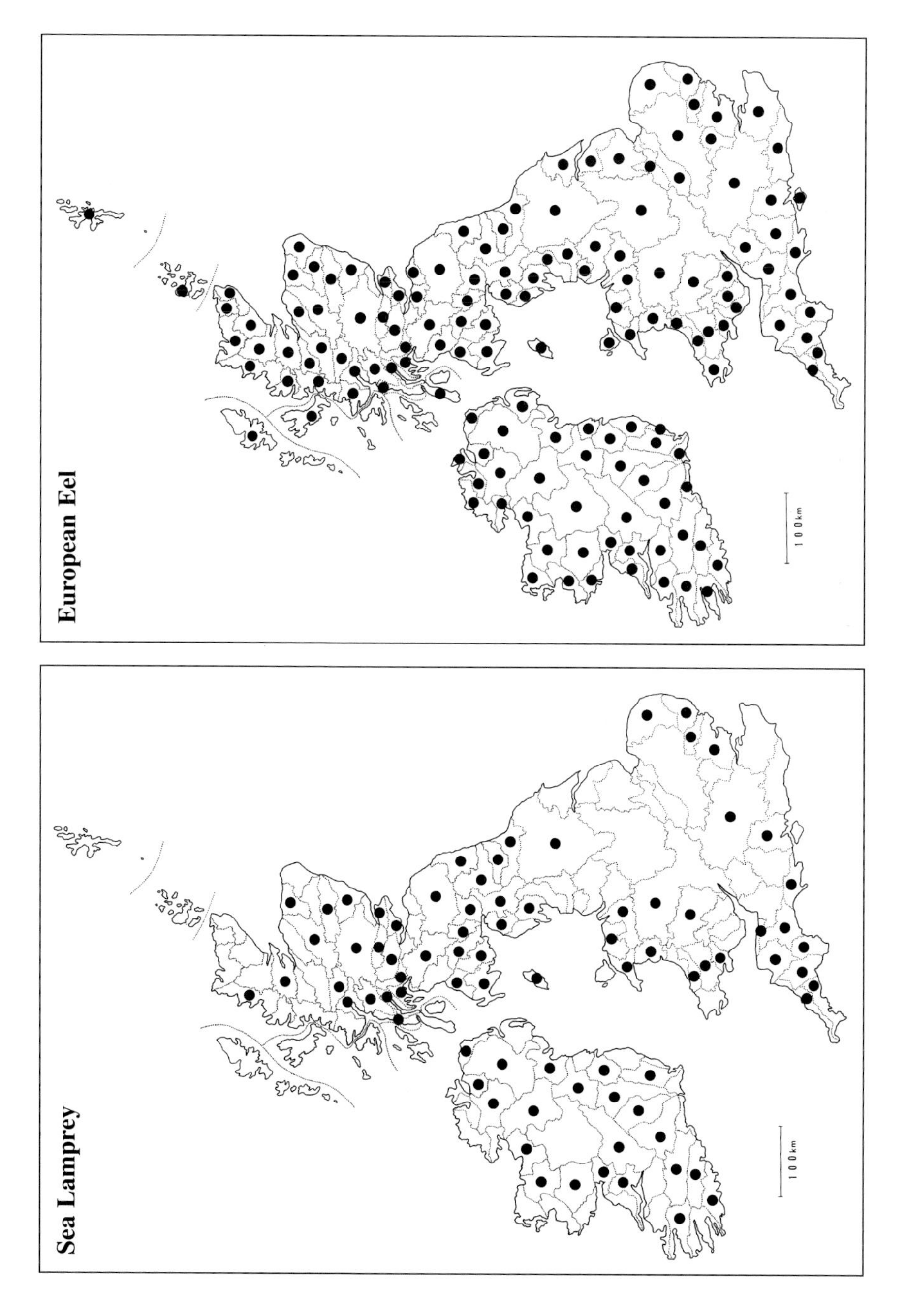
European Eel
100km
Sea Lamprey
100km

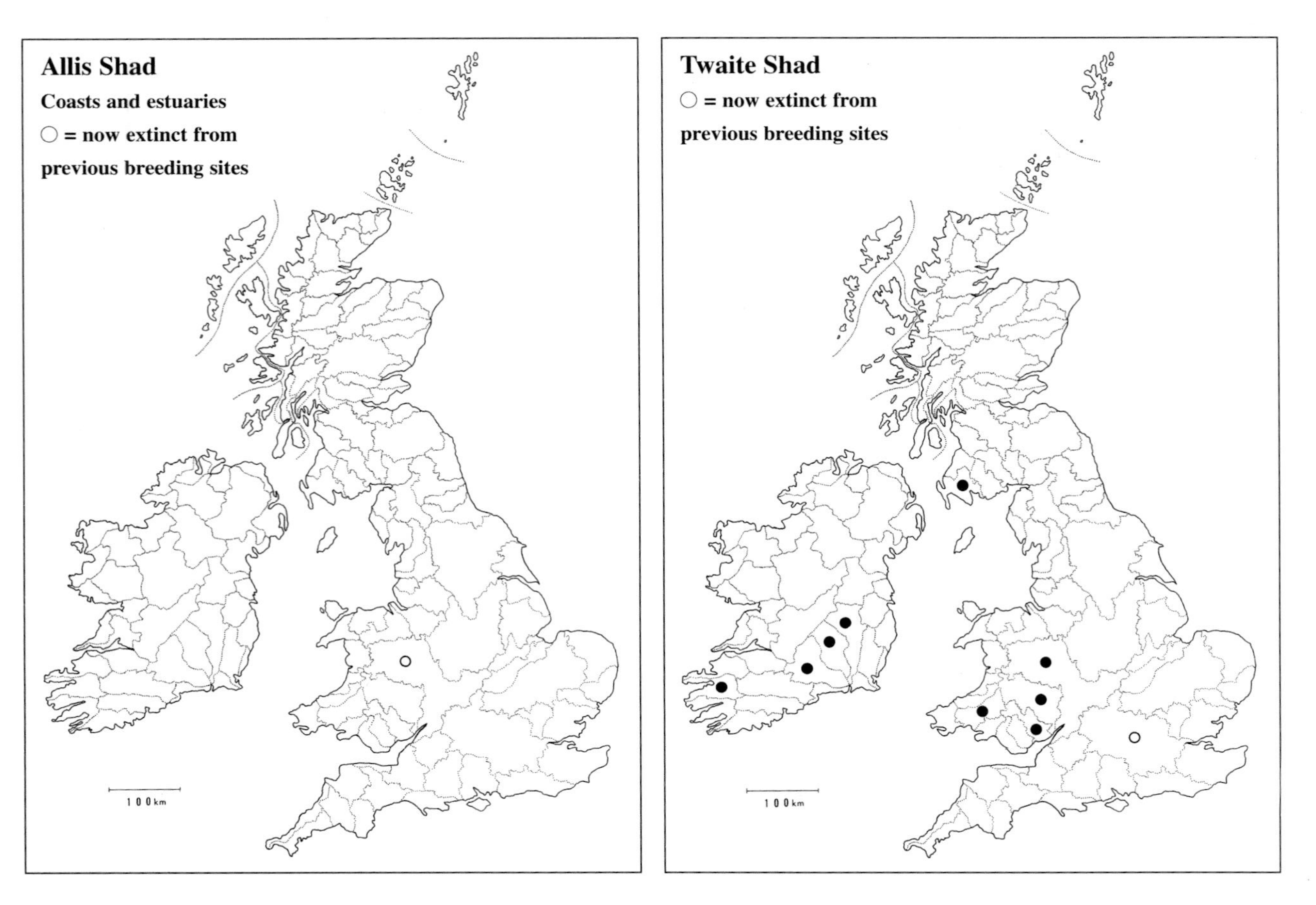
Allis Shad
Coasts and estuaries
○ = now extinct from
previous breeding sites
100 km
Twaite Shad
○ = now extinct from
previous breeding sites
100 km

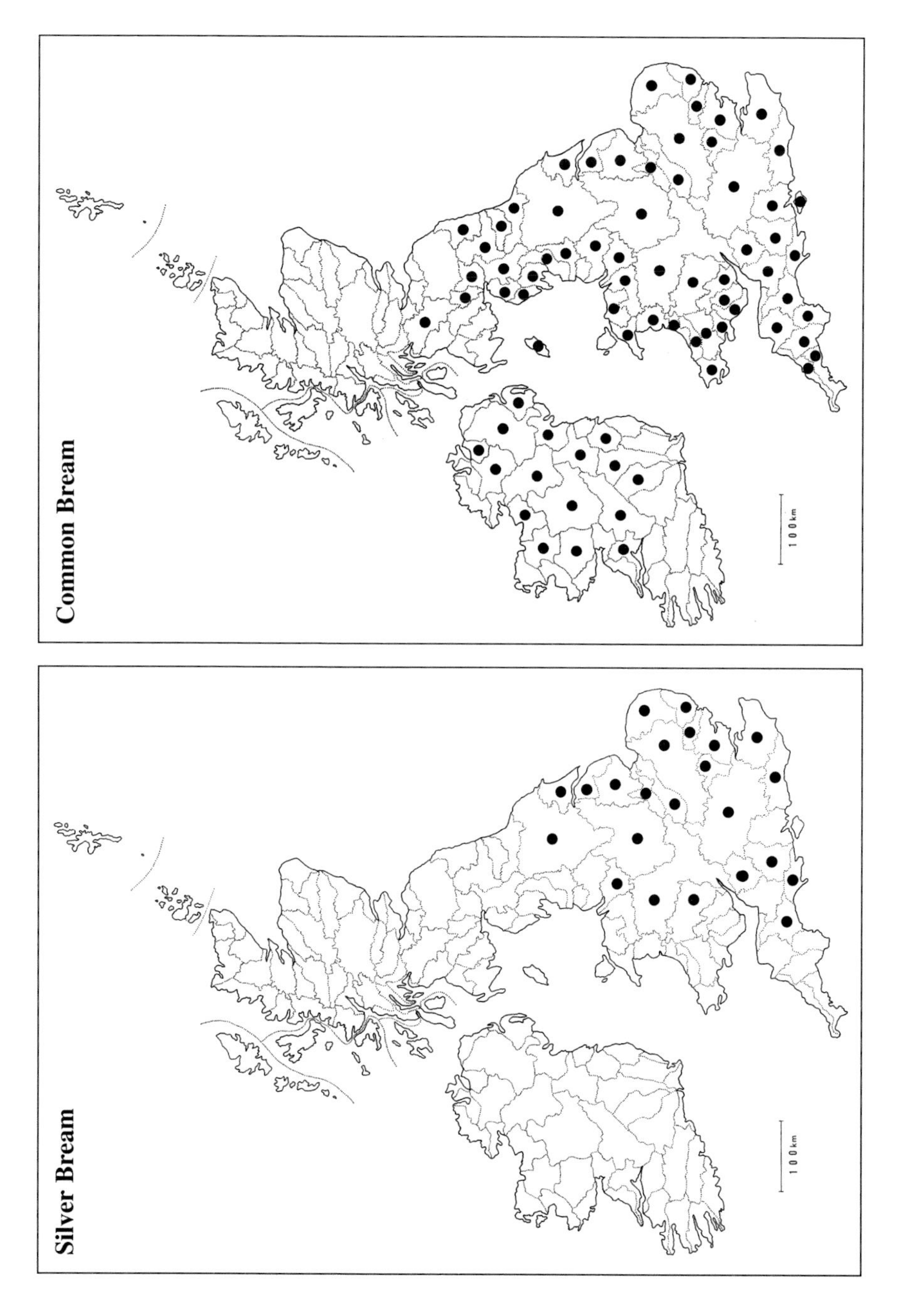
Common Bream
100km
Silver Bream
100km

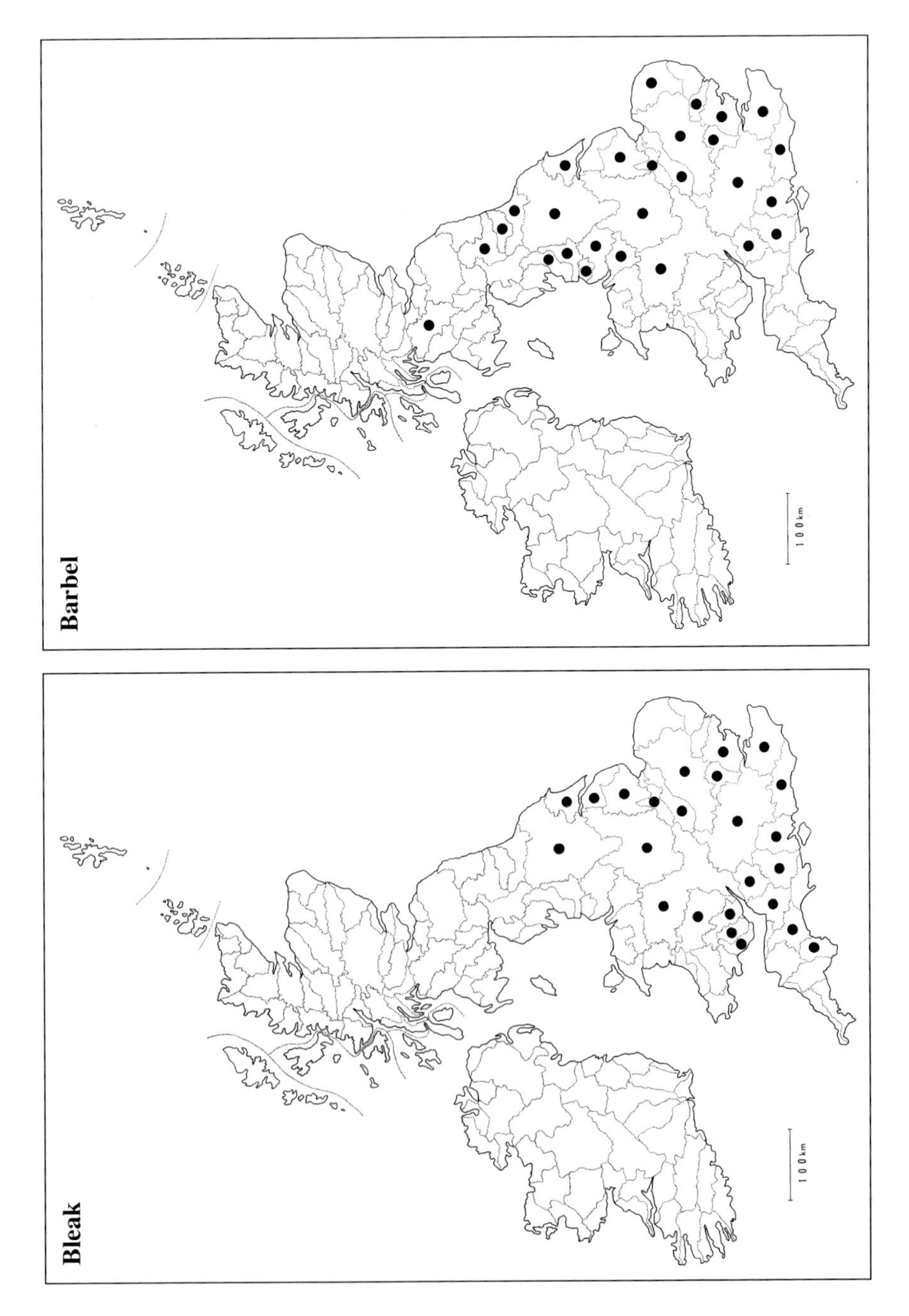
Barbel
100 km
Bleak
100 km

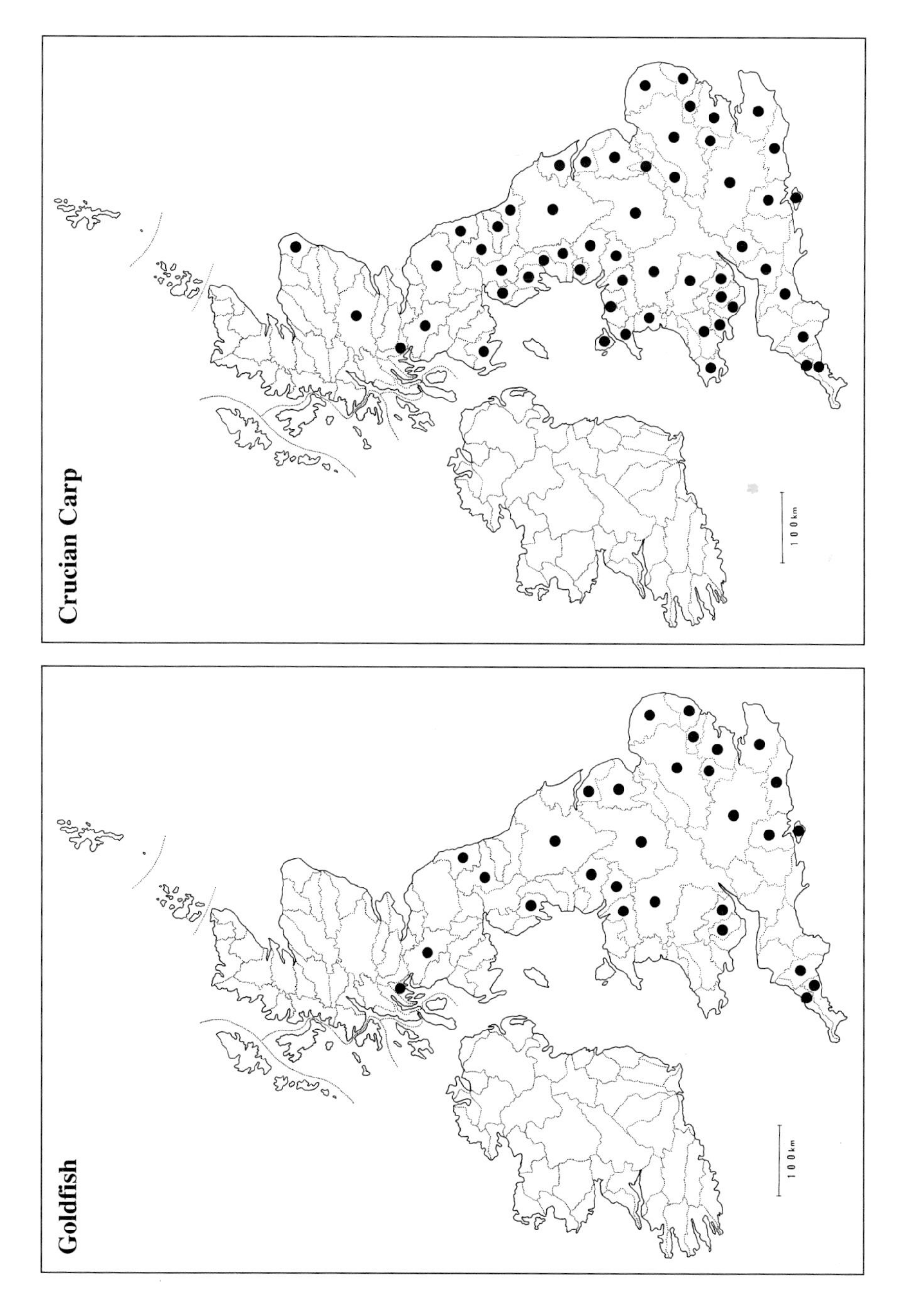
Crucian Carp
100 km
Goldfish
100 km

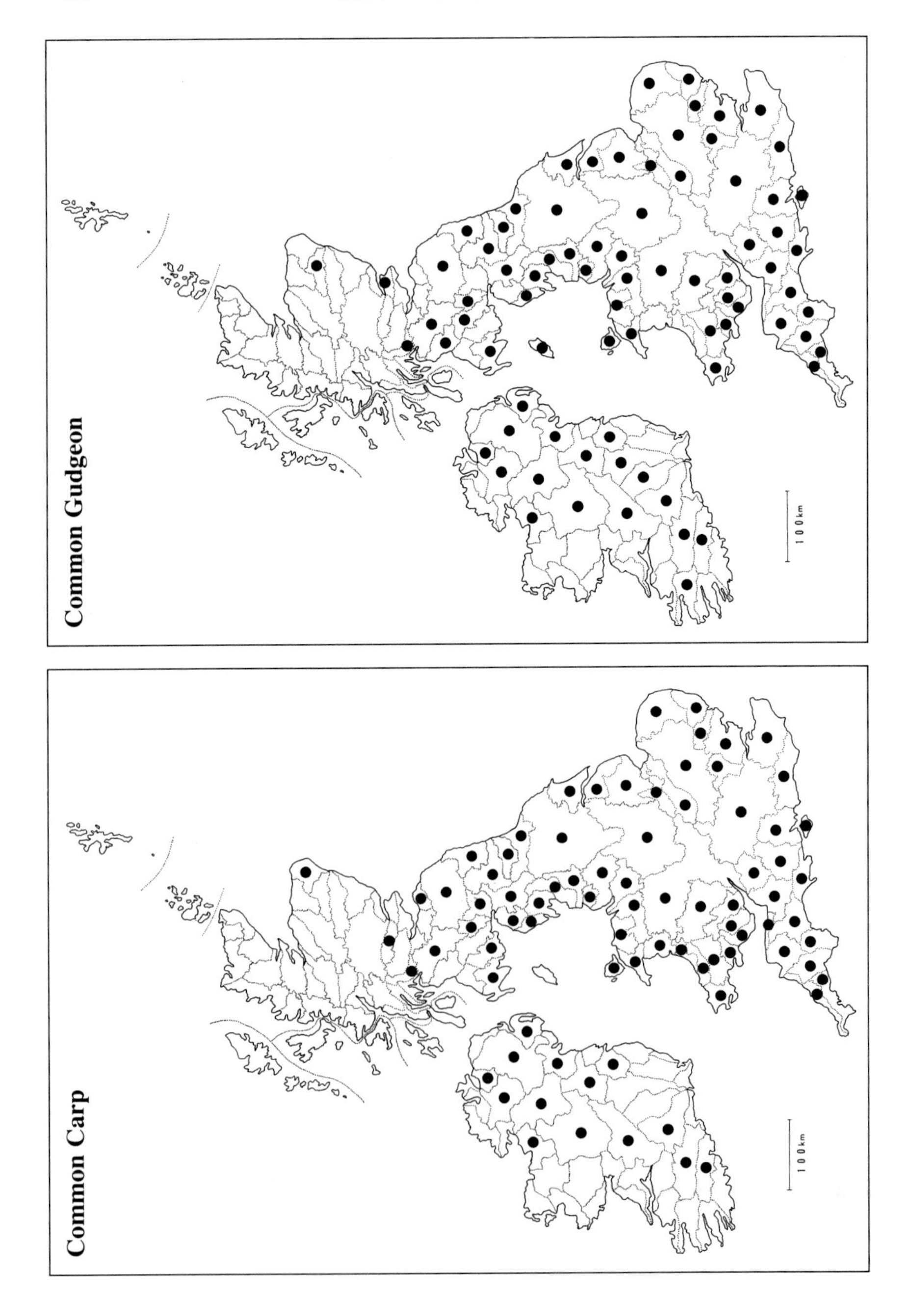
Common Gudgeon
100km
Common Carp
100km

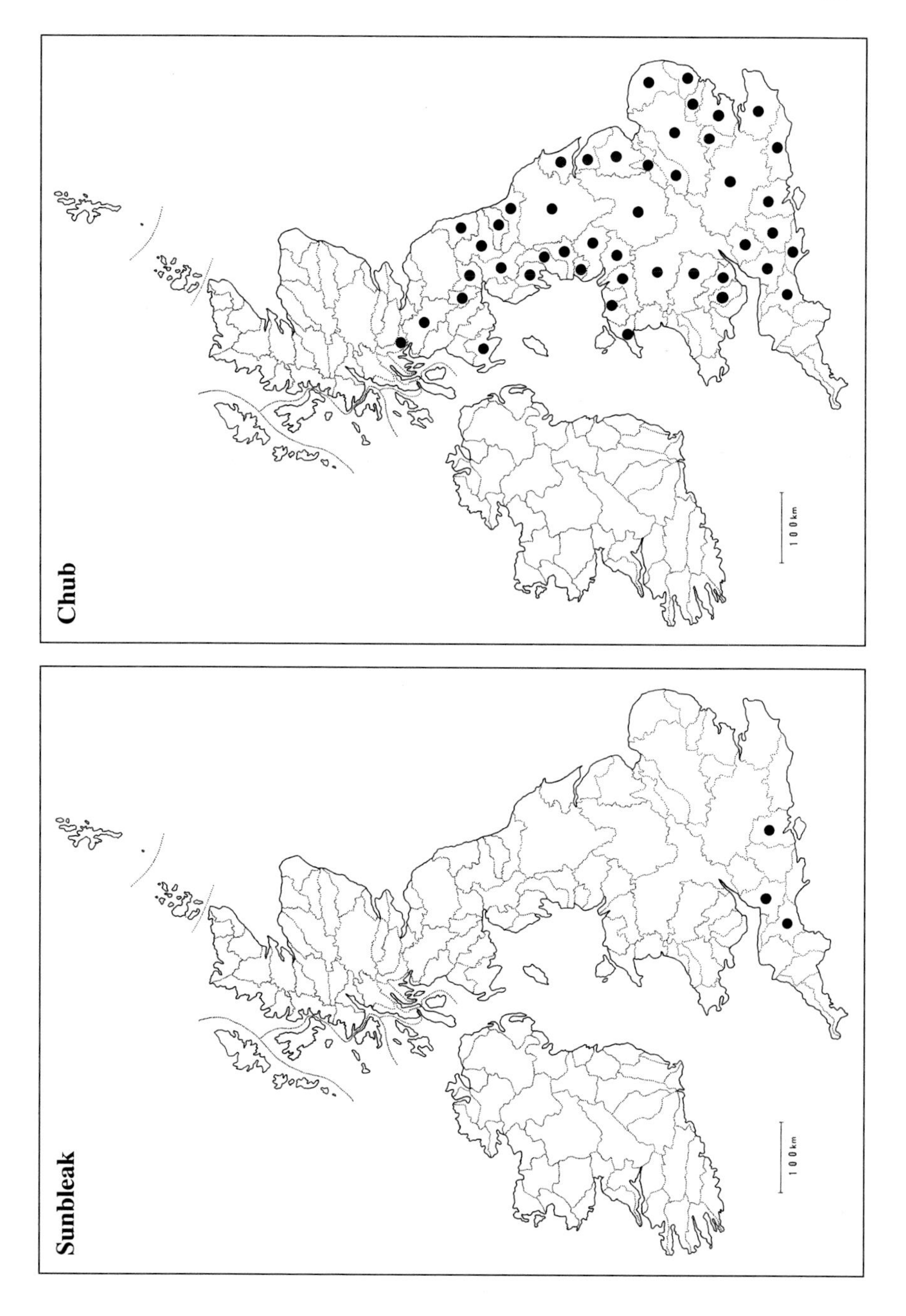
Chub
100km
Sunbleak
100km

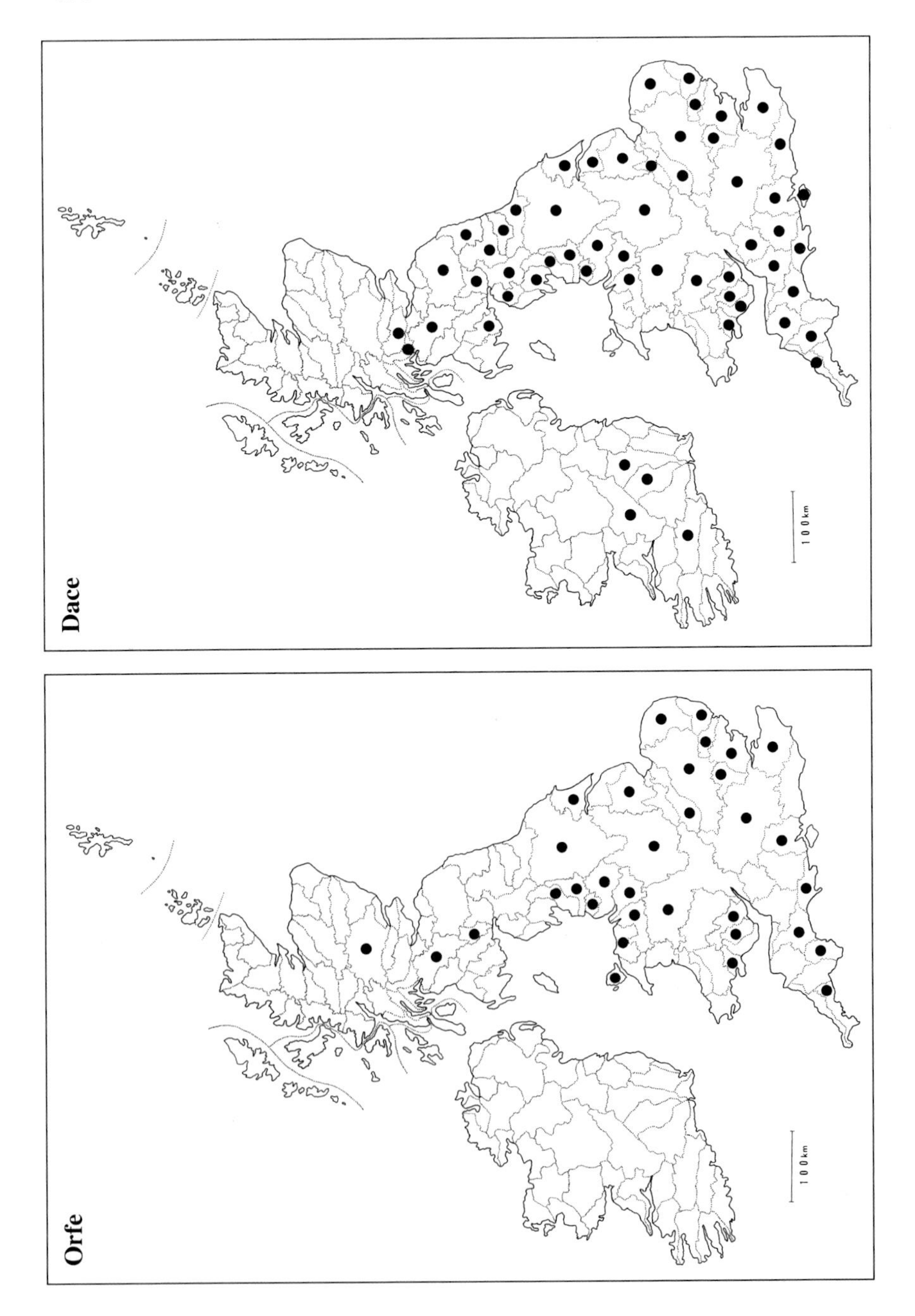
Dace
100km
Orfe
100km

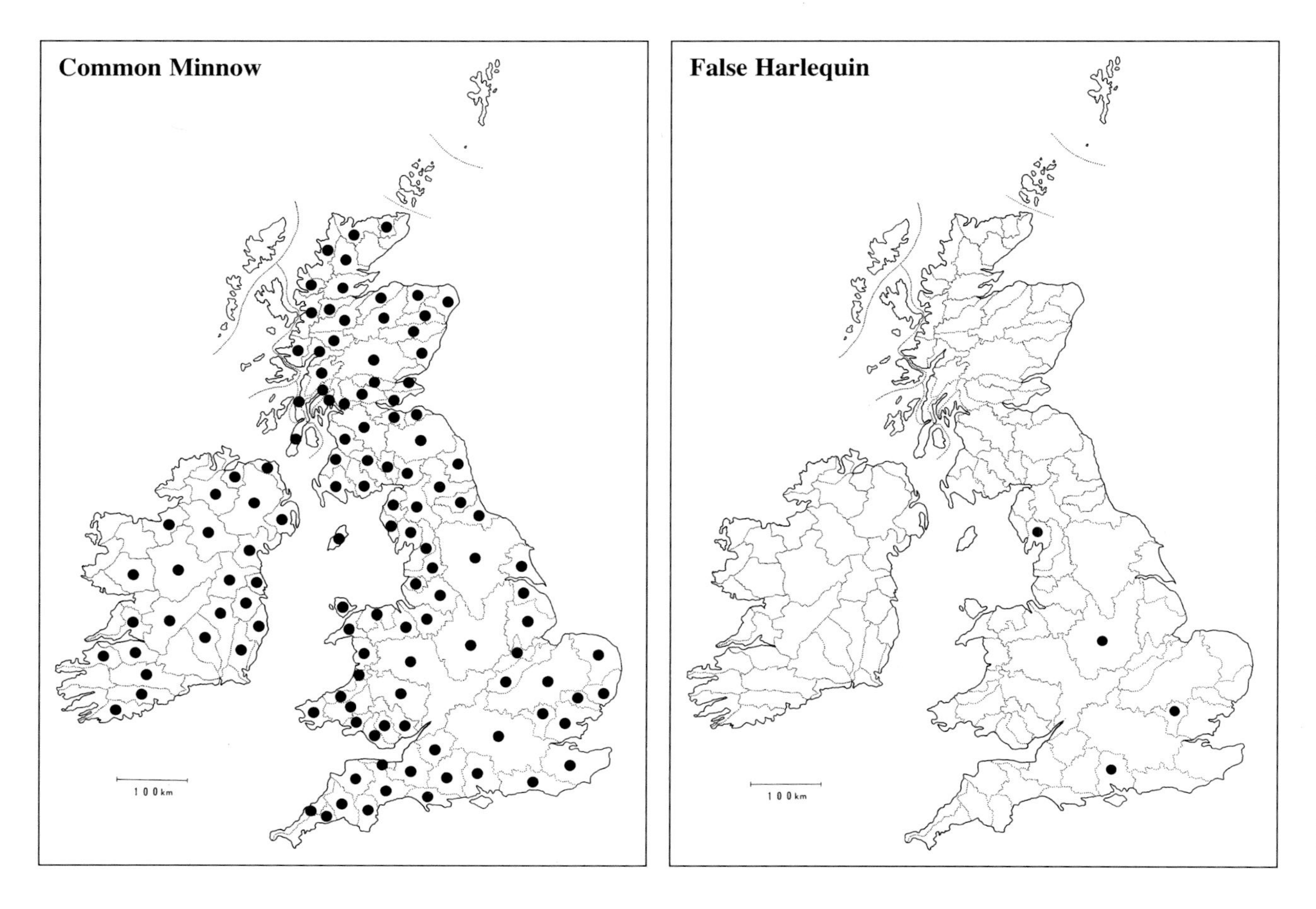
Common Minnow
100 km
False Harlequin
100 km

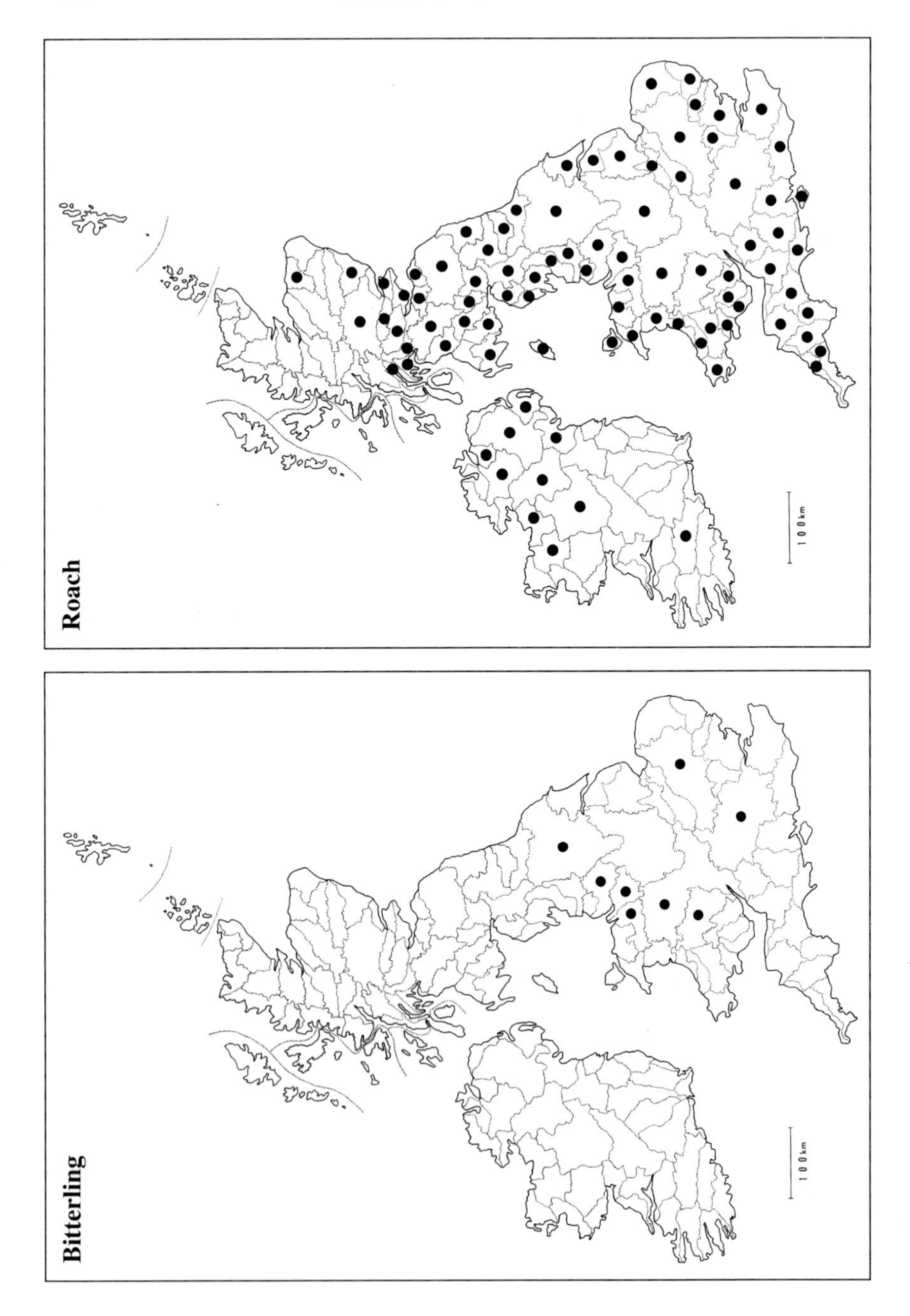
Roach
100 km
Bitterling
100 km

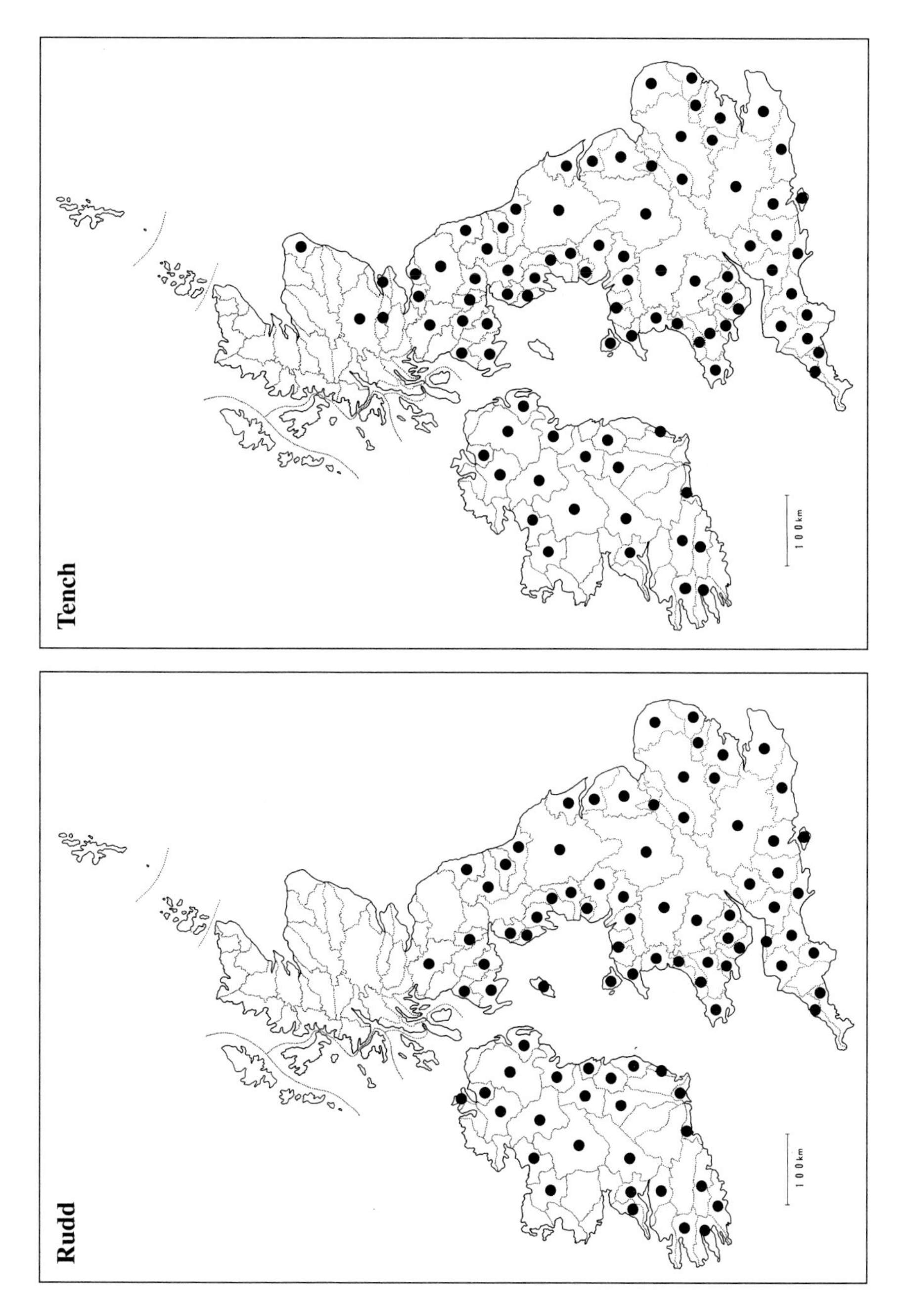
Tench
100 km
Rudd
100 km

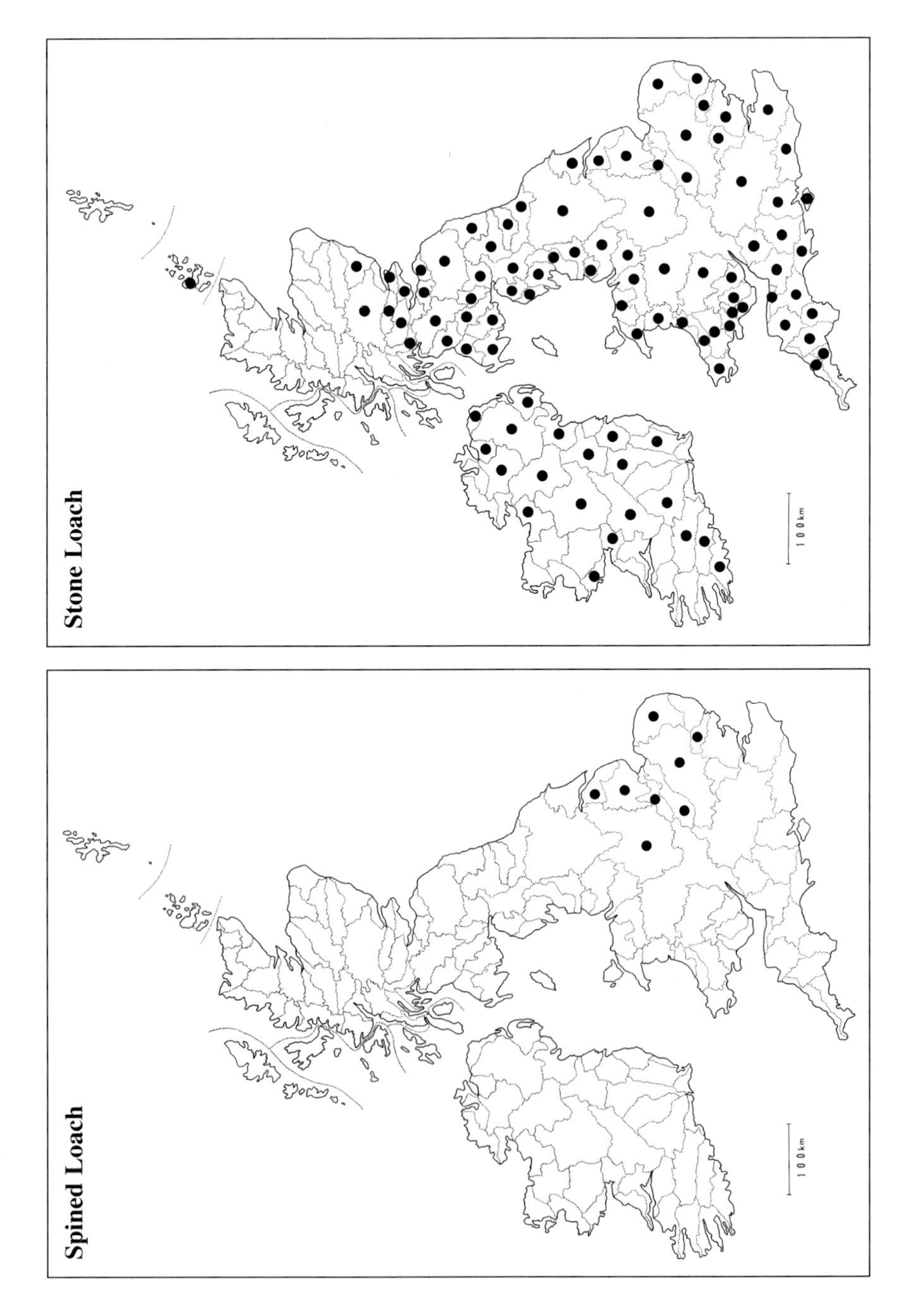
Stone Loach
100km
Spined Loach
100km

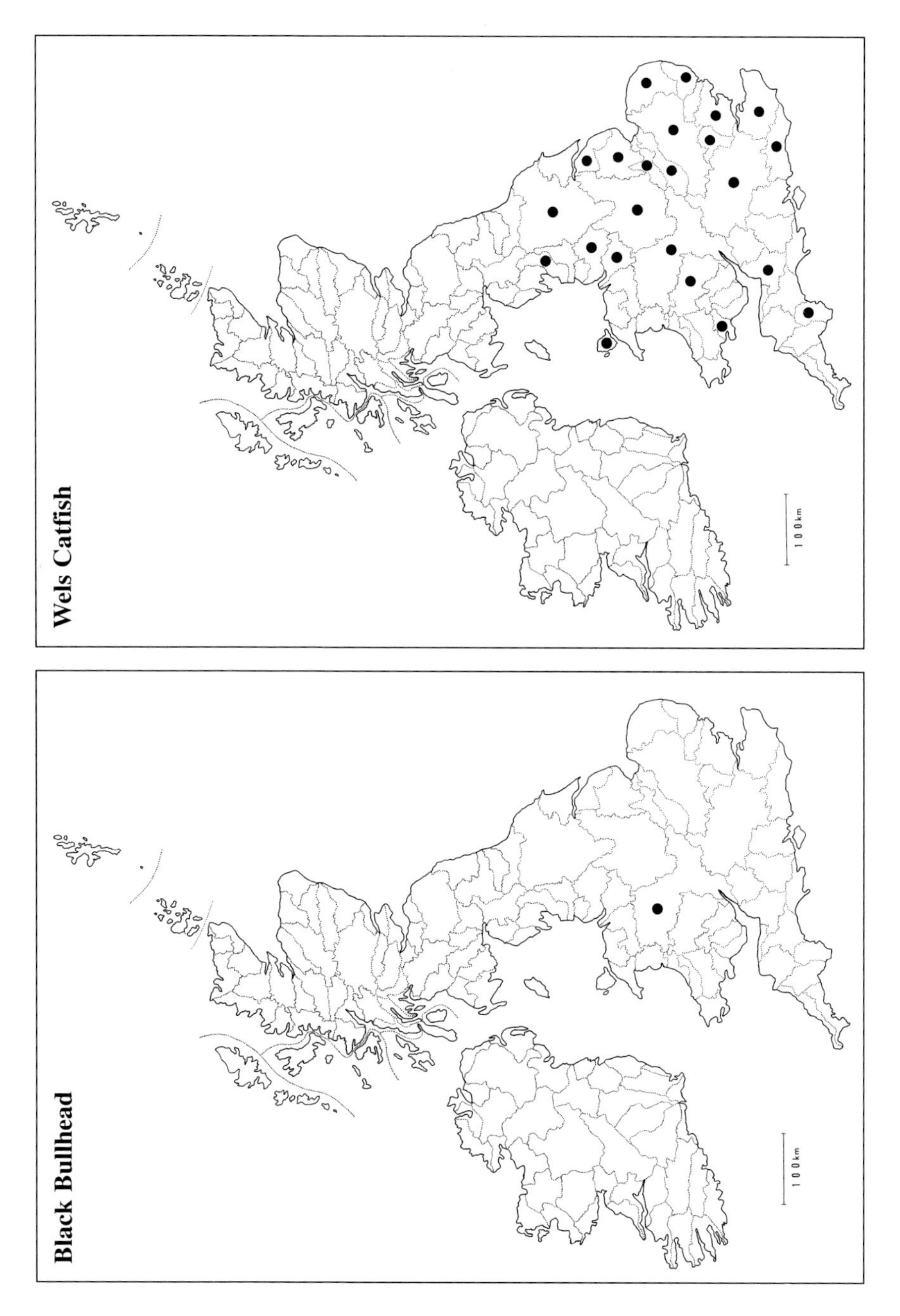
Wels Catfish
100 km
Black Bullhead
100 km

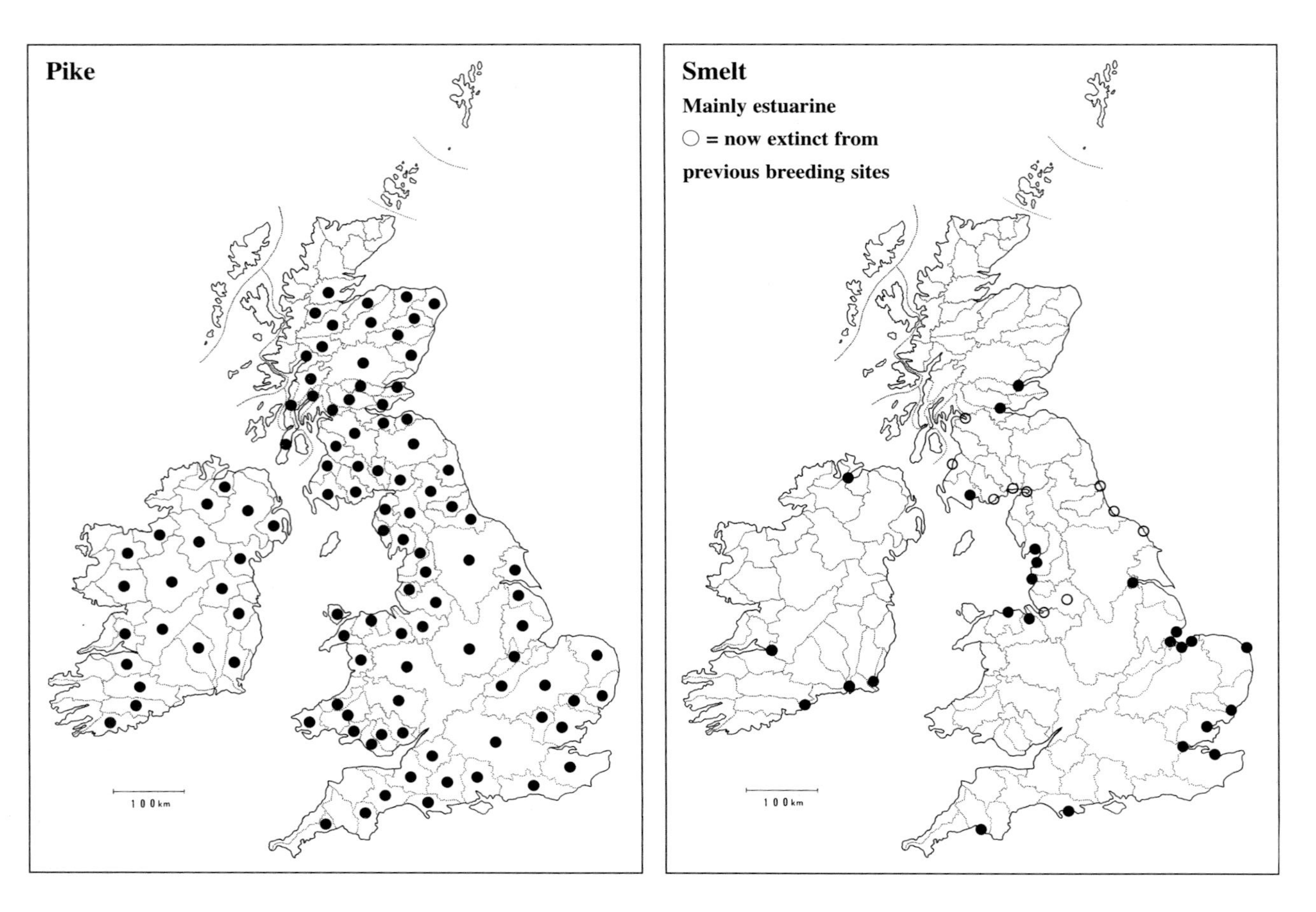
Pike
100km
Smelt
Mainly estuarine
○ = now extinct from
previous breeding sites
100km

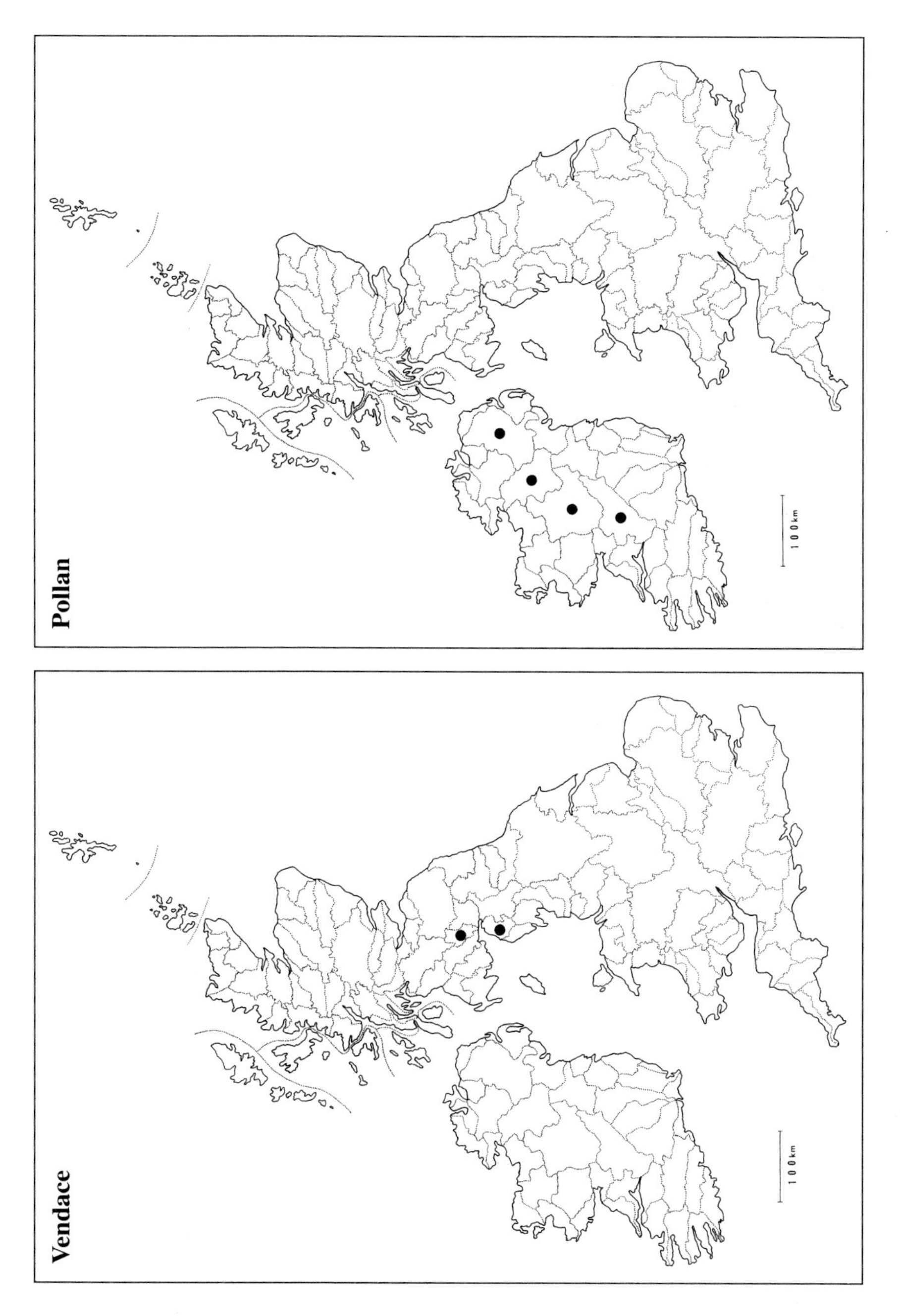
Pollan
100km
Vendace
100km

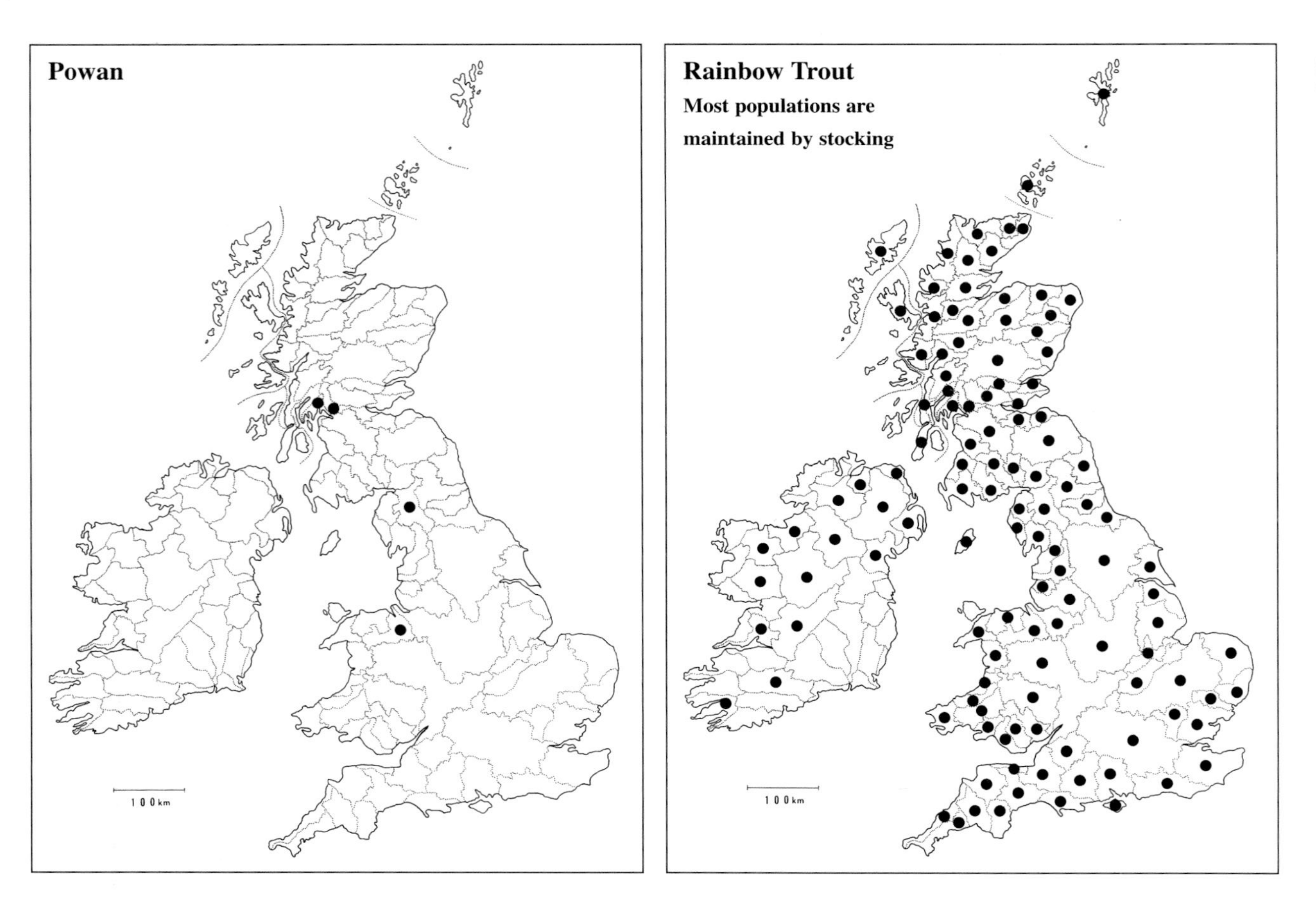
Powan
100 km
Rainbow Trout
Most populations are
maintained by stocking
100 km

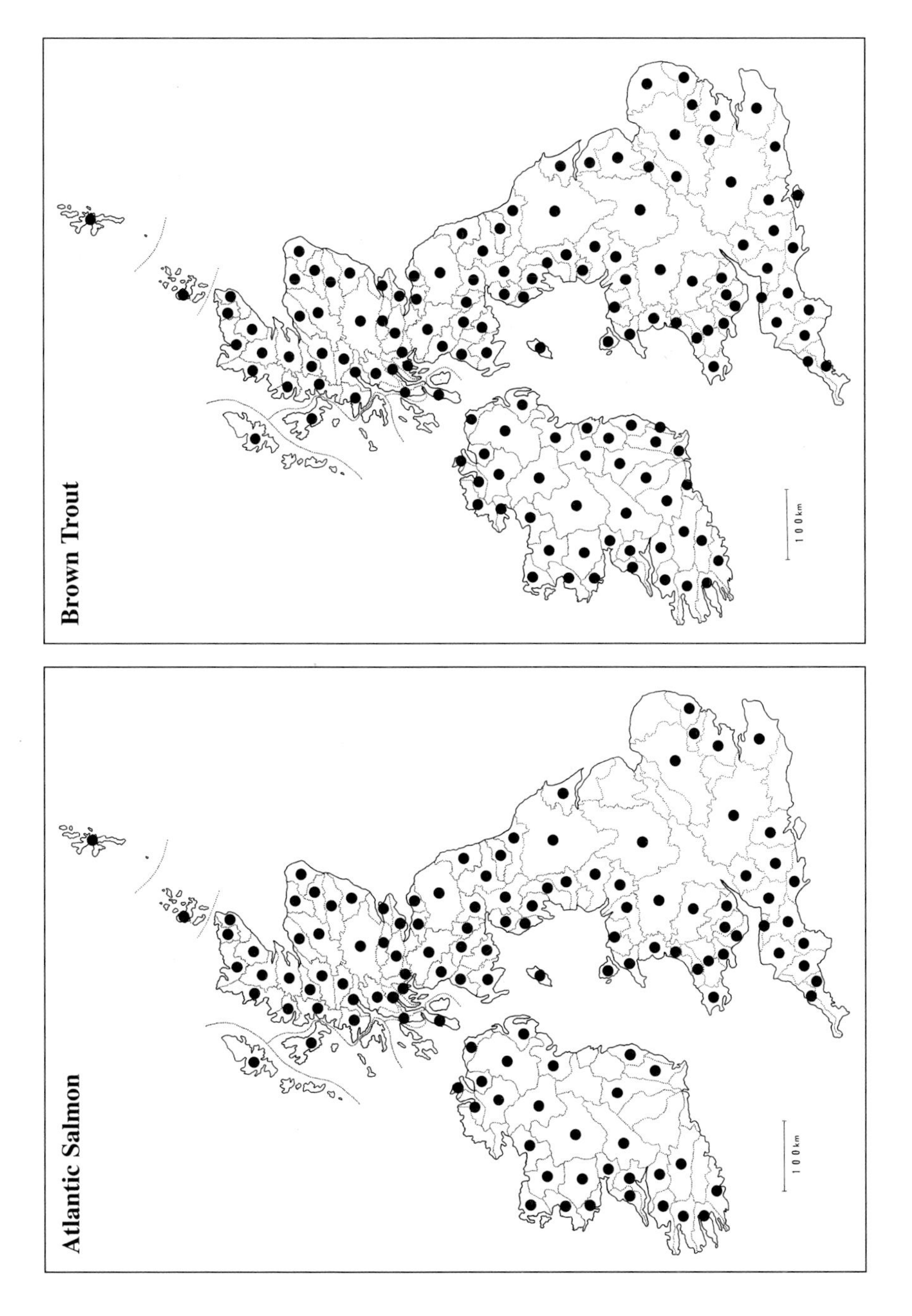
Brown Trout
100km
Atlantic Salmon
100km

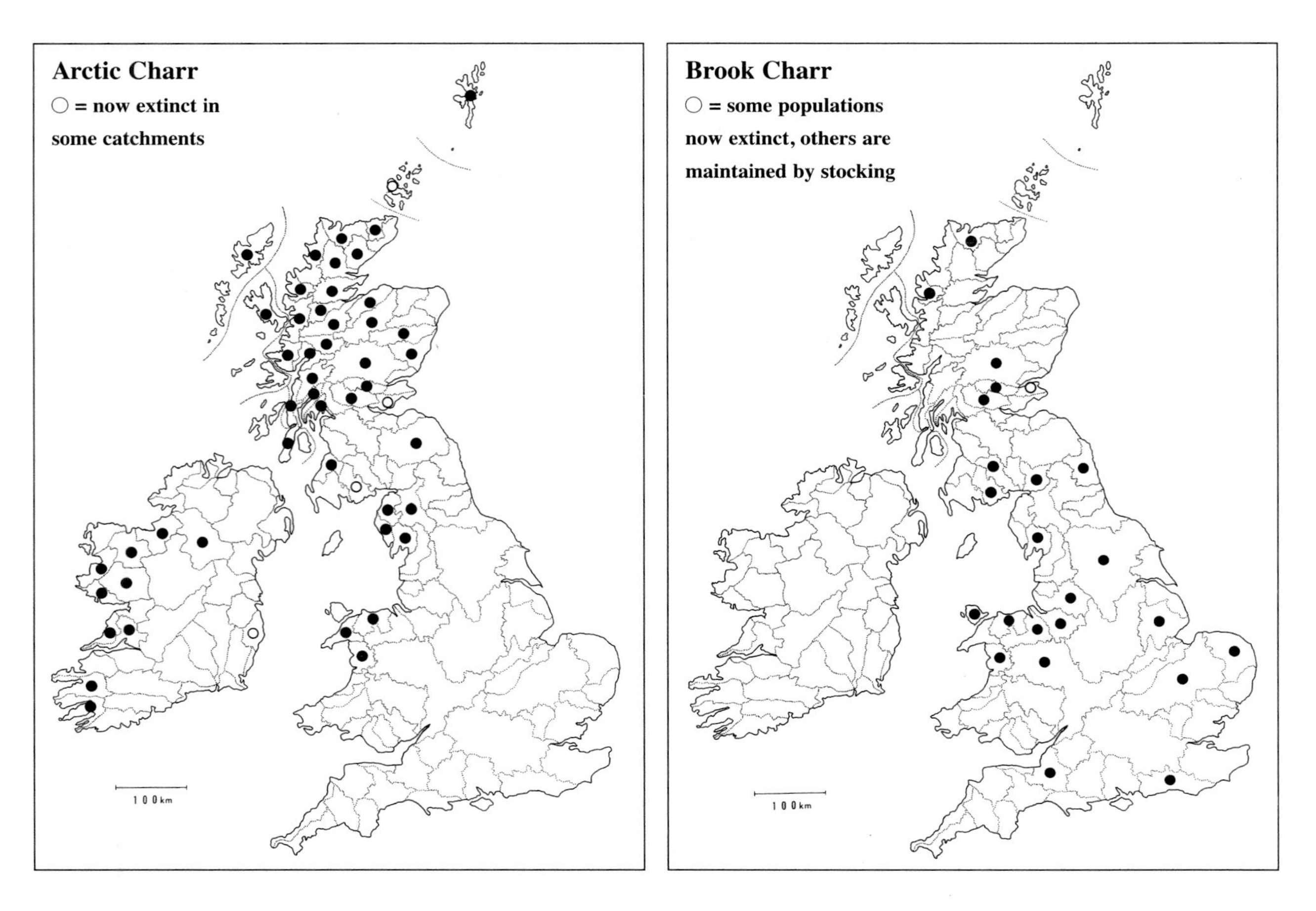
Arctic Charr
○ = now extinct in
some catchments
100 km
Brook Charr
○ = some populations
now extinct, others are
maintained by stocking
100 km

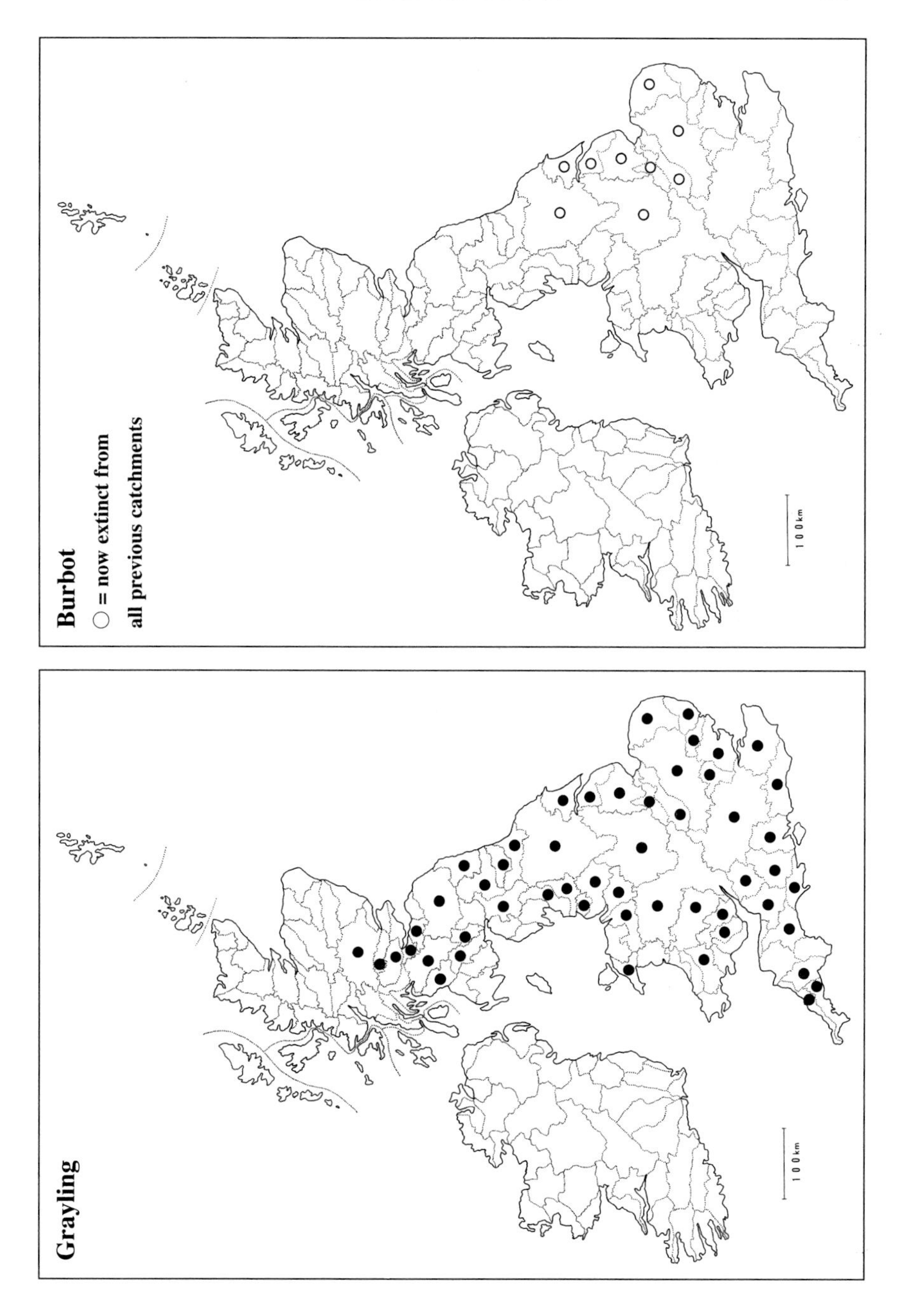
Burbot
○ = now extinct from
all previous catchments
100km
Grayling
100km

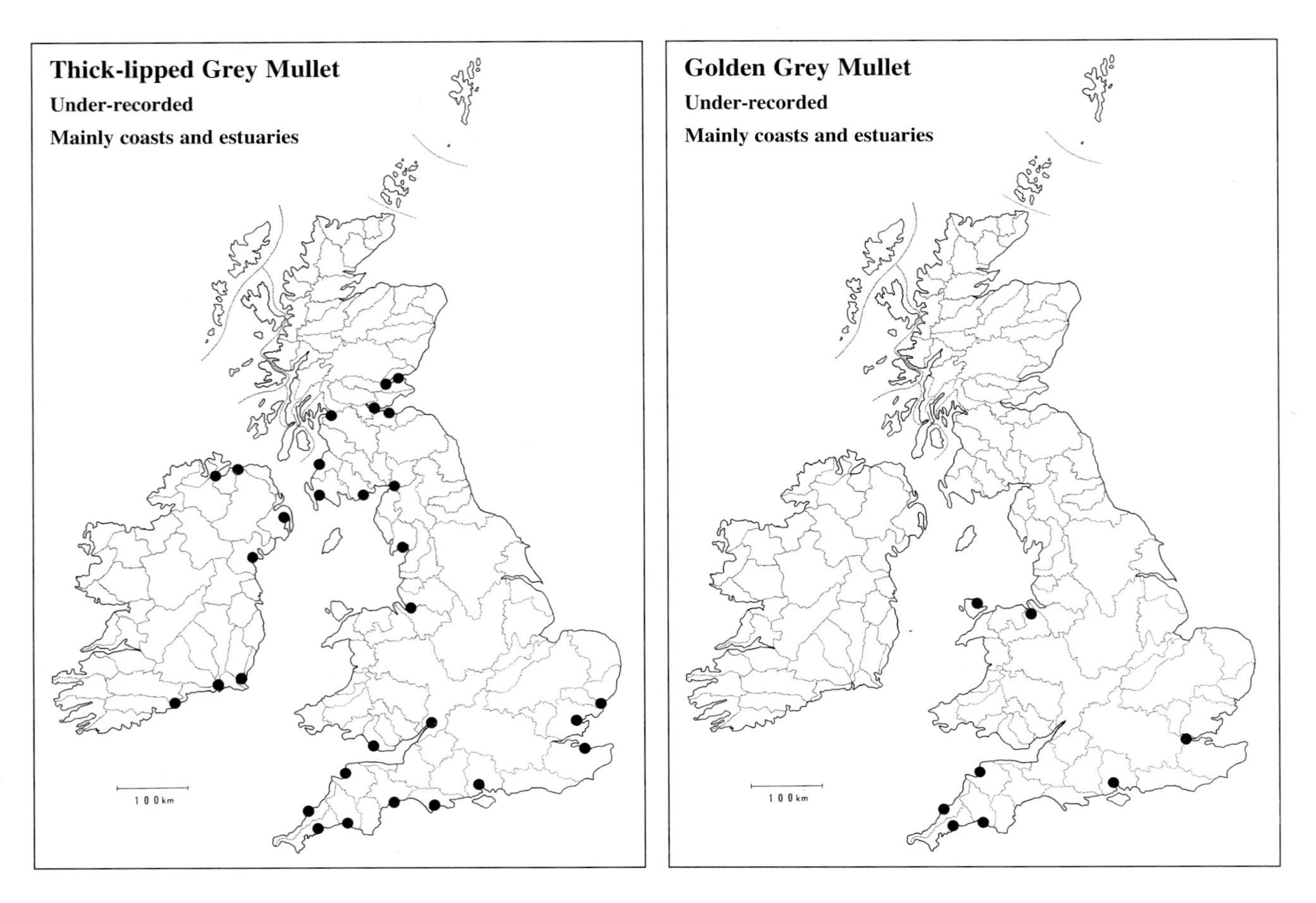
Thick-lipped Grey Mullet
Under-recorded
Mainly coasts and estuaries
100km
Golden Grey Mullet
Under-recorded
Mainly coasts and estuaries
100km

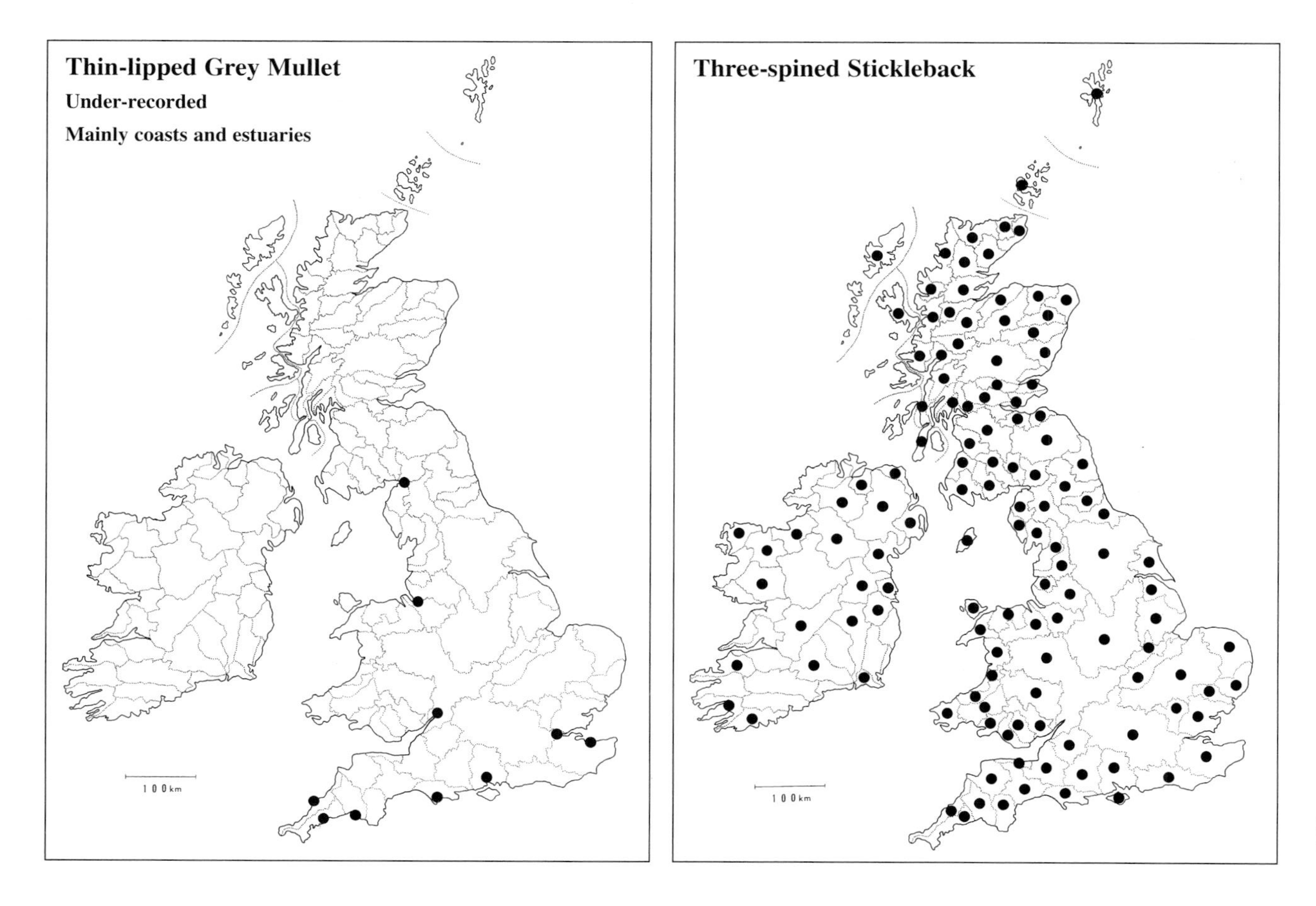
Thin-lipped Grey Mullet
Under-recorded
Mainly coasts and estuaries
100 km
Three-spined Stickleback
100 km

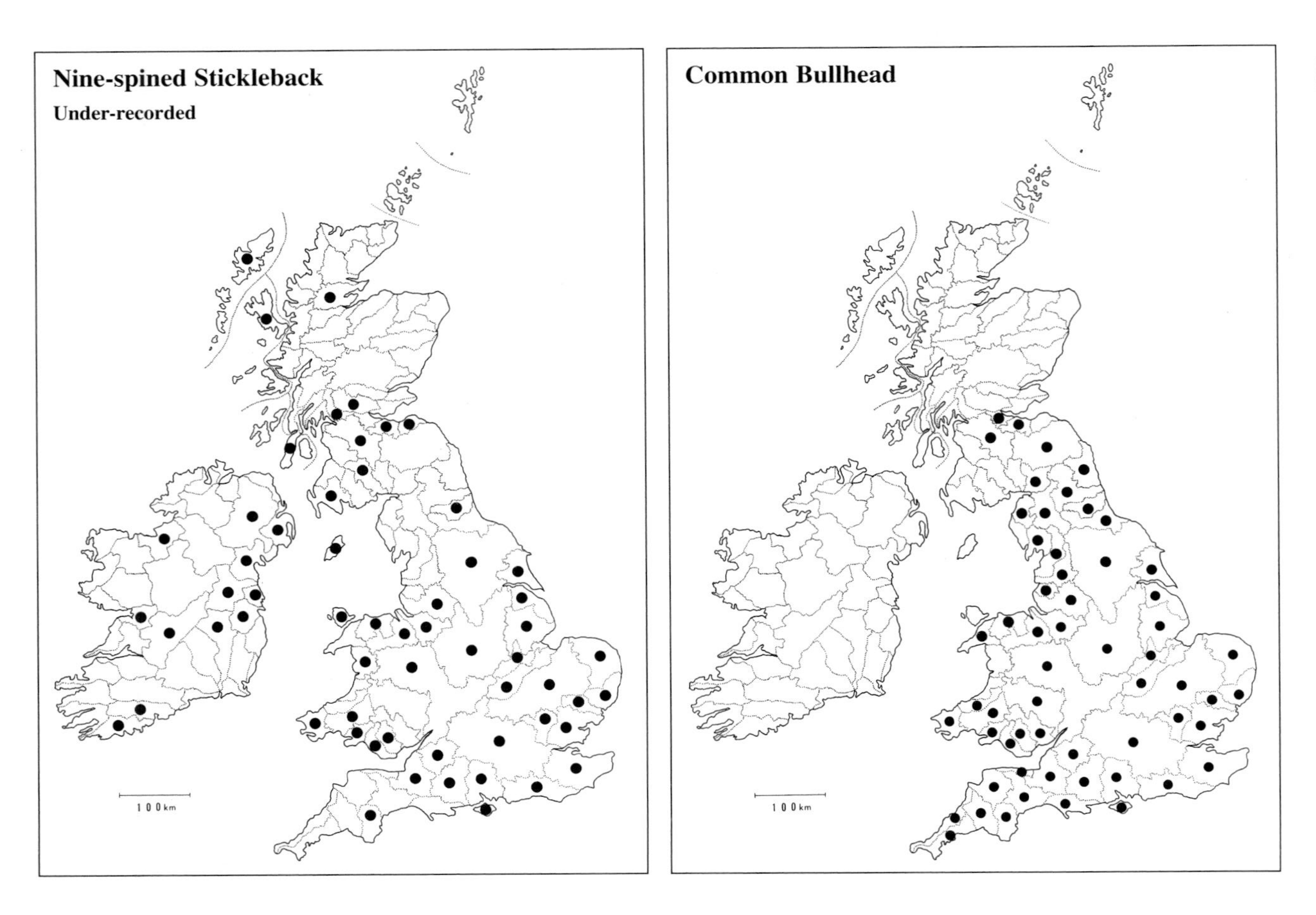
Nine-spined Stickleback
Under-recorded
100 km
Common Bullhead
100 km

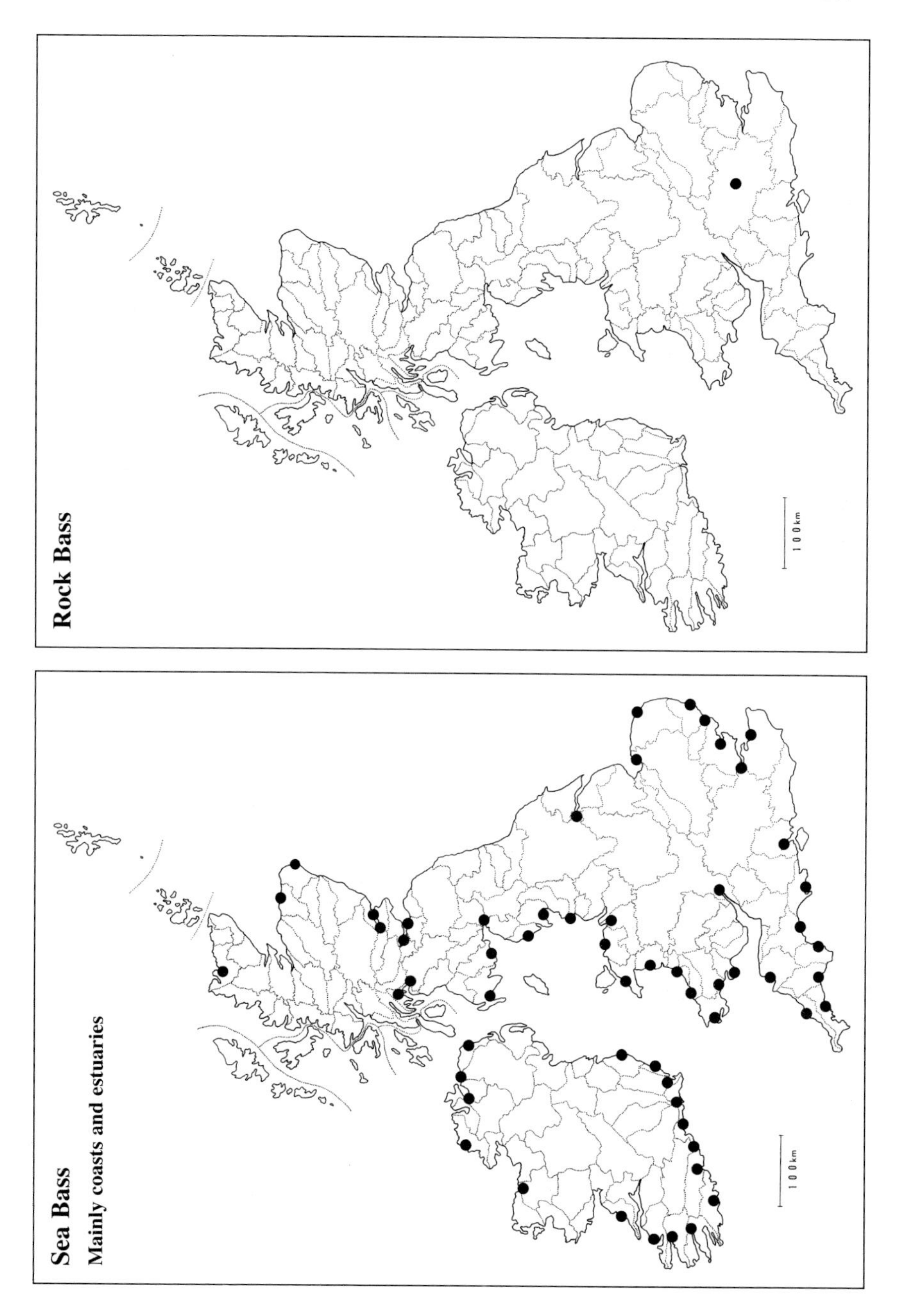
Rock Bass
100 km
Sea Bass
Mainly coasts and estuaries
100 km

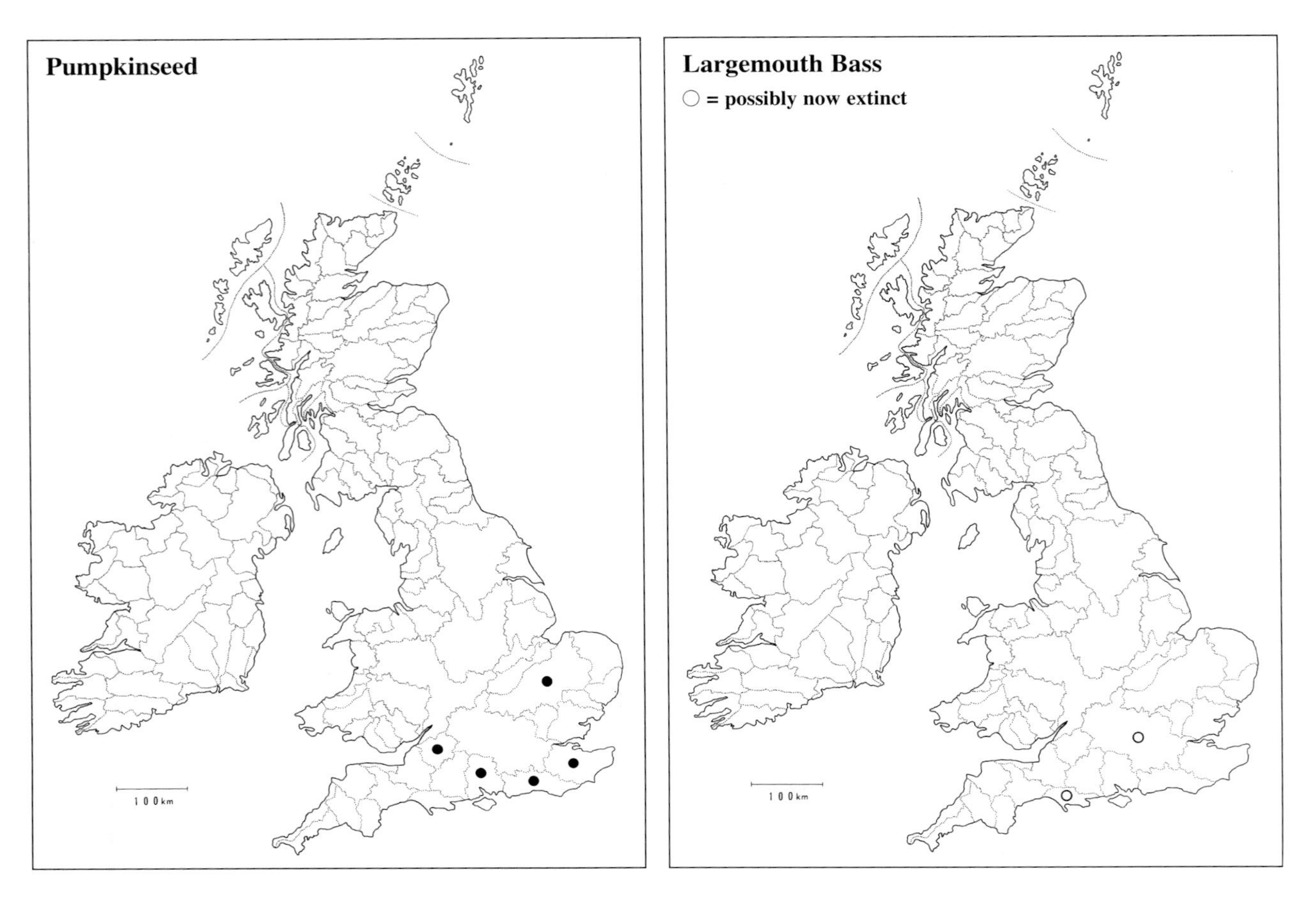
Pumpkinseed
100km
Largemouth Bass
○ = possibly now extinct
100km

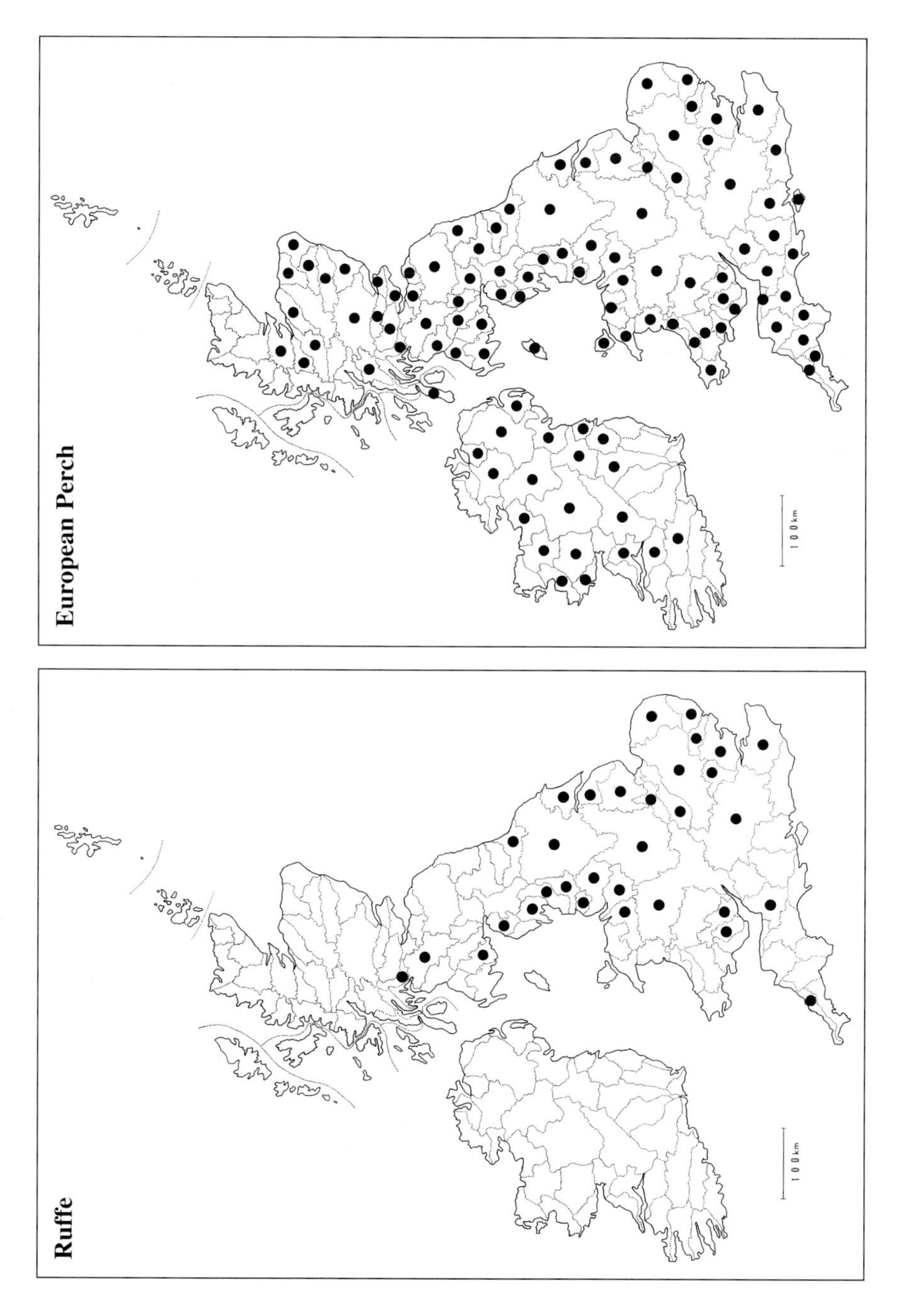
European Perch
100 km
Ruffe
100 km

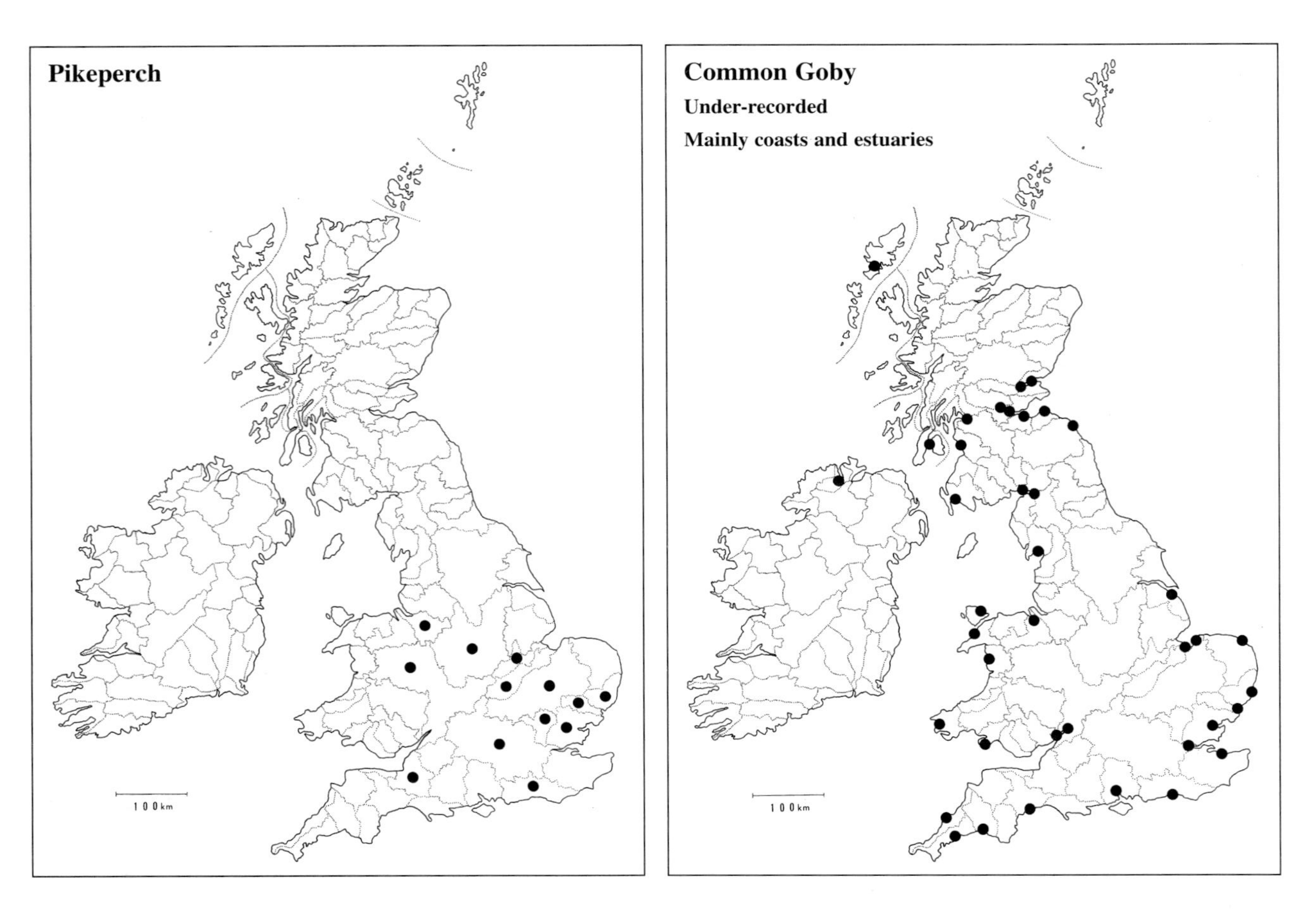
Pikeperch
100 km
Common Goby
Under-recorded
Mainly coasts and estuaries
100 km

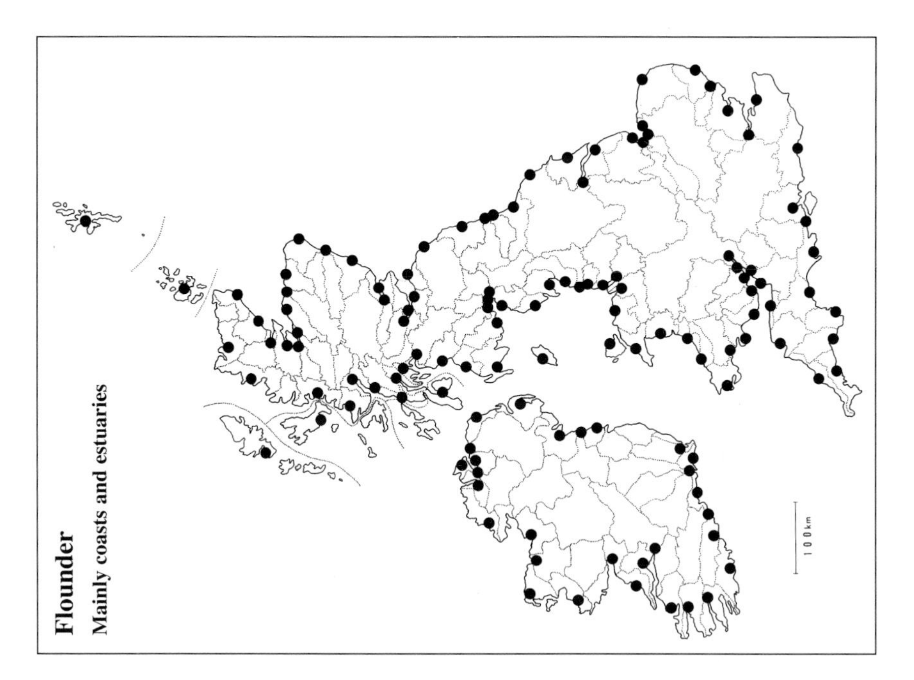
Flounder
Mainly coasts and estuaries
100 km

ACKNOWLEDGEMENTS

I am extremely grateful to David Sutcliffe for his care and patience in editing this publication. Helpful remarks on an earlier draft were received from Phil Hickley, Roger Sweeting and Ian Winfield. I thank the artists – Deitrich Burkel, Joanna Langhorne, Keith Linsell and Gordon Reid – who have joined with me in illustrating the keys, and also the photographers who have supplied photographs. Chris Gardner was particularly helpful in supplying photographs from the Environment Agency, and Rob Britton kindly took photographs of scales from Flounder. Jane Robins helped with references, Adrian Pinder with information on new alien species in England, and Joe Caffrey, Fran Igoe, Jimmy King, Julia Nunn and Robert Rosell provided Irish data. I am indebted to these and many other colleagues who have supplied useful comments and information since the first key was published in 1972.

Aass, P. (1972). Age determination and year-class fluctuations of Cisco, *Coregonus albula* L., in the Mjosa hydroelectric reservoir, Norway. *Report of the Institute of Freshwater Research, Drottningholm* 52, 5-22. **Abakumov, V.A. (1964).** On the systematics and ecology of the brook lamprey. *Voprosy Ikhtiologii* 34, 423-430. **Abou-Seedo, F.S. & Potter, I.C. (1979).** The estuarine phase in the spawning run of the river lamprey *Lampetra fluviatilis*. *Journal of Zoology, London* 188, 5-25. **Adams, C.E., Brown, D.W. & Tippett, R. (1990).** Dace (*Leuciscus leuciscus* (L.)) and Chub (*Leuciscus cephalus* (L.)) new introductions to the Loch Lomond catchment. *Glasgow Naturalist* 21, 509-513. **Adams, C.E., Fraser, D., Huntingford, F.A., Greer, R.B., Askew, C.M. & Walker, A.F. (1998).** Trophic polymorphism amongst Arctic charr from Loch Rannoch, Scotland. *Journal of Fish Biology* 52, 1259-1271. **Adams, C.E. & Maitland, P.S. (1998).** The Ruffe population of Loch Lomond, Scotland: its introduction, population expansion and interaction with native species. *Journal of Great Lakes Research* 24, 249-262. **Adams, C.E. & Maitland, P.S. (2002).** Invasion and establishment of freshwater fish populations in Scotland – the experience of the past and lessons for the future. *Glasgow Naturalist* 23, 35-43. **Adams, C.E. & Mitchell, J. (1992).** Introduction of another non-native fish species to Loch Lomond: Crucian Carp (*Carassius carassius* (L.)). *Glasgow Naturalist* 22, 165-168. **Adams, C.E. & Tippett, R. (1991).** Powan ova (*Coregonus lavaretus*) predation by introduced Ruffe (*Gymnocephalus cernuus*) in Loch Lomond, Scotland. *Aquaculture and Fisheries Management* 22, 261-267. **Aldridge, D.C. (1999).** Development of European bitterling in the gills of freshwater mussels. *Journal of Fish Biology* 54, 138-151. **Al-Hassan, L.A., Webb, C.J., Giama, M. & Miller, P.J. (1987).** Phosphoglucose isomerase polymorphism in the common goby, *Pomatoschistus microps* (Kroyer) (Teleostei: Gobiidae), around the British Isles. *Journal of Fish Biology* 30, 281-298. **Ali, S.S. (1976).** The food of Roach, *Rutilus rutilus* (L.) in Llyn Tegid (North Wales). *Sind University Research Journal of Science* 9, 15-33. **Ali, S.S. (1979).** Age, growth and length-weight relationship of the Roach *Rutilus rutilus* L. in Llyn Tegid, north Wales. *Pakistan Journal of Zoology* 11, 1-19. **Allardi, J. & Chancerel, F. (1988).** Note ichtyologique sur la presence en France de *Pseudorasbora parva* (Schlegel 1842). *Bulletin Francais de Peche et Pisciculture* 308, 35-37. **Allen, A. (1987).** The olfactory world of the goldfish. *Aquarist and Pondkeeper* 1987, 52. **Allen, J.R.M. & Wootton, R.J. (1982).** Age, growth and rate of food consumption in an upland population of the Three-spined Stickleback, *Gasterosteus aculeatus*. *Journal of Fish Biology* 21, 95-106. **Allen, K.R. (1938).** Some observations of the biology of the Trout (*Salmo trutta*) in Windermere. *Journal of Animal Ecology* 7, 333-349. **Almaca, C. (1988).** On the Sturgeon, *Acipenser sturio*, in the Portuguese rivers and sea. *Folia Zoologica* 37, 183-191. **Almeida, P.R., Moreira, F., Costa, J.L., Assis, C.A. & Costa, M.J. (1993).** The feeding strategies of *Liza ramada* (Risso 1826) in fresh and brackish water in the River Tagus, Portugal. *Journal of Fish Biology* 42, 95-107. **Anders, K. & Weise, V. (1993).** Glochidia of the freshwater mussel, *Anodonta anatina*, affects the anadromous European smelt (*Osmerus eperlanus*) from the Eider estuary, Germany. *Journal of Fish Biology* 42, 411-420. **Anderson, M. (1982).** The identification of British grey mullets. *Journal of Fish*

Biology 20, 33-38. **Andrews, C. (1979).** Host specificity of the parasite fauna of Perch (*Perca fluviatilis* L.) from the British Isles, with special reference to the study at Llyn Tegid. *Journal of Fish Biology* 15, 195-209. **Andrews, C.W. & Lear, E. (1956).** The biology of Arctic char (*Salvelinus alpinus* L.) in northern Labrador. *Journal of the Fisheries Research Board of Canada* 13, 843-860. **Applegate (1950).** Natural history of the Sea Lamprey, *Petromyzon marinus*, in Michigan. *Special Scientific Report, U.S. Fish and Wildlife Service* 55, 1-237. **Aprahamian, M.W. (1981).** Aspects of the biology of the twaite shad (*Alosa fallax*) in the Rivers Severn and Wye. *Proceedings of the Second British Coarse Fish Conference, Liverpool* 2, 373-381. **Aprahamian, M.W. (1985).** The effect of the migration of *Alosa fallax fallax* (Lacepede) into fresh water, on branchial and gut parasites. *Journal of Fish Biology* 27, 521-532. **Aprahamian, M.W. (1988).** The biology of the Twaite Shad *Alosa fallax fallax* (Lacepede) in the Severn Estuary. *Journal of Fish Biology* 33A, 141-152. **Aprahamian, M.W. (1989).** The diet of juvenile and adult Twaite Shad *Alosa fallax fallax* (Lacepede) from the Rivers Severn and Wye (Britain). *Hydrobiologia* 179, 173-182. **Aprahamian, M.W., Jones, G.O. & Gough, P.J. (1998).** Movement of Atlantic salmon in the Usk estuary, Wales. *Journal of Fish Biology* 53, 221-225. **Araujo, F.G., Bailey, R.G. & Williams, W.P. (1999).** Spatial and temporal variations in fish populations in the upper Thames Estuary. *Journal of Fish Biology* 55, 836-853. **Assis, C.A., Almeida, P.R., Moreira, F. Costa, J.L. & Costa, M.J. (1992).** Diet of the Twaite Shad *Alosa fallax* (Lacepede) (Clupeidae) in the River Tagus estuary, Portugal. *Journal of Fish Biology* 41, 1049-1050. **Ausen, V. (1976).** Age, growth, population size, mortality and yield in the Whitefish (*Coregonus lavaretus* (L.)) of Haugatjern – a eutrophic Norwegian lake. *Norwegian Journal of Zoology* 24, 379-405.

Backe-Hansen, P. (1982). Age determination, growth and maturity of the Bleak *Alburnus alburnus* (L.) (Cyprinidae) in Lake Oyeren, SE Norway. *Fauna Norvegica* 3, 31-36. **Bagenal, T.B. (1969a).** The relationship between food supply and fecundity in Brown Trout *Salmo trutta* L. *Journal of Fish Biology* 1, 167-182. **Bagenal, T.B. (1969b).** Relationship between egg size and fry survival in Brown Trout *Salmo trutta* L. *Journal of Fish Biology* 1, 349-353. **Bagenal, T.B. (1970).** Notes on the biology of the Schelly *Coregonus lavaretus* (L.) in Haweswater and Ullswater. *Journal of Fish Biology* 2, 137-154. **Bagenal, T.B. (1973).** *Identification of British Fishes.* Hulton, Amersham. **Bagenal, T.B. (Ed.) (1978).** *Methods for Assessment of Fish Production in Freshwaters*. IBP Handbook No. 3. Blackwell, Oxford. **Bailey, M.M. (1972).** Age, growth, reproduction and food of the Burbot, *Lota lota* (Linnaeus), in southwestern Lake Superior. *Transactions of the American Fisheries Society* 101, 667-674. **Balinsky, B.I. (1948).** On the development of specific characters in cyprinid fishes. *Proceedings of the Zoological Society, London* 118, 335-344. **Ball, J.N. & Jones, J.W. (1960).** On the growth of Brown Trout of Llyn Tegid. *Proceedings of the Zoological Society of London* 134, 1-41. **Banks, J. (1970).** Observations on the fish population of Rostherne Mere, Cheshire. *Field Studies* 3, 375-379. **Barannikova, I.A. (1987).** Review of sturgeon farming in the Soviet Union. *Journal of Ichthyology* C, 62-71. **Baras, E. & Philippart, J.C. (1999).** Adaptive and evolutionary significance of a reproductive thermal threshold in *Barbus barbus*. *Journal of Fish Biology* 55, 354-375. **Barbour, S.E. &**

Einarsson, S.M. (1987). Ageing and growth of Charr *Salvelinus alpinus* (L.) from habitat types in Scotland. *Aquaculture and Fisheries Management* 18, 1-13. **Bardonnet, A. & Gaudin, P. (1990).** Diel pattern of first downstream post-emergence displacement in grayling, *Thymallus thymallus* (L., 1758). *Journal of Fish Biology* 37, 623-628. **Bardonnet, A., Gaudin, P. & Thorpe, J.E. (1993).** Diel rhythm of emergence and of first displacement downstream in trout (*Salmo trutta*), Atlantic salmon (*Salmo salar*) and grayling (*Thymallus thymallus*). *Journal of Fish Biology* 43, 755-762. **Bastl, I. (1977).** Notes on reproduction biology of the bleak – *Alburnus alburnus* (Linnaeus 1758) – in the Vojka system of Danube arms, Czechoslovakia. *Biologia* 32, 591-598. **Baxter, E.W. (1954).** Lamprey distribution in streams and rivers. *Nature, London* 180, 1145. **Beamish, F.W.H. & Medland, T.E. (1988).** Age determination for lampreys. *Transactions of the American Fisheries Society* 113, 63-71. **Bean, C.W. & Winfield, I.J. (1989).** Biological and ecological effects of a *Ligula intestinalis* (L.) infestation of the gudgeon, *Gobio gobio* (L.), in Lough Neagh, Northern Ireland. *Journal of Fish Biology* 34, 135-148. **Bean, C.W. & Winfield, I.J. (1992).** Influences of the tapeworm *Ligula intestinalis* (L.) on the spatial distributions of juvenile roach *Rutilus rutilus* (L.) and gudgeon *Gobio gobio* (L.) in Lough Neagh, Northern Ireland. *Netherlands Journal of Zoology* 42, 416-429. **Bean, C.W. & Winfield, I.J. (1995).** Habitat use and activity patterns of roach (*Rutilus rutilus* (L.)), rudd (*Scardinius erythrophthalmus* (L.)), perch (*Perca fluviatilis* L.) and pike (*Esox lucius* L.) in the laboratory: the role of predation threat and structural complexity. *Ecology of Freshwater Fish* 4, 37-46. **Bean, C.W., Winfield, I.J. & Fletcher, J.M. (1996).** Stock assessment of the Arctic charr (*Salvelinus alpinus*) population in Loch Ness, U.K. Pages 206-223 in: Cowx, I.G. (Ed.) *Stock Assessment in Inland Fisheries*. Fishing News Books, Oxford. **Beaumont, A.R., Bray, J., Murphy, J.M. & Winfield, I.J. (1995).** Genetics of whitefish and vendace in England and Wales. *Journal of Fish Biology* 46, 880-890. **Beaumont, W.R.C. & Mann, R.H.K. (1983).** The age, growth and diet of a freshwater population of the Flounder, *Platichthys flesus* (L.), in southern England. *Journal of Fish Biology* 23, 607-616. **Becker, G.C. (1983).** *Fishes of Wisconsin*. University of Wisconsin Press, Madison. **Behnke, R.J. (1984).** Steelhead trout. *Trout* 25, 43-48. **Belaud, A., Cassou-Leins, F., Cassou-Leins, J.J. & Labat, R. (1991).** La ponte d'un poisson migrateur de la Garonee la grande alose (*Alosa alosa* L.). *Ichtyophysiologica Acta* 14, 123-126. **Belaud, A., Dautrey, R., Labat, R., Lartigue, J.P. & Lim, P. (1985).** Observations sur le comportement migratoires des Aloses (*Alosa alosa* L.) dans le canal artificiel de l'usine de Golfech. *Annales Limnologiques* 21, 161-172. **Belyanina, T.N. (1969).** Synopsis of biological data on smelt, *Osmerus eperlanus* (Linnaeus). *Food and Agriculture Organisation Fishery Synopsis, Rome* 78, 1-55. **Bembo, D.G., Beverton, R.J.H., Weightman, A.J. & Cresswell, R.C. (1993).** Distribution, growth and movement of River Usk brown trout (*Salmo trutta*). *Journal of Fish Biology* 43A, 45-52. **Berg, L.S. (1965).** *Freshwater Fishes of the U.S.S.R and Adjacent Countries*. Israel Program for Scientific Translations, Jerusalem. **Bertin, L. (1956).** *Eels*. Cleaver-Hume, London. **Bettoli, P.W., Neill, W.H. & Kelsch, S.W. (1985).** Temperature preference and heat resistance of grass carp, *Ctenopharyngodon idella* (Valenciennes), bighead carp, *Hypophthalmichthys nobilis* (Gray) and their F1 hybrid. *Journal of Fish Biology* 27, 239-248. **Beukema, J.J. (1970).** Acquired hook avoidance in pike, *Esox lucius* L., fished with artificial and natural baits. *Journal of Fish Biology* 2, 155-160. **Beukema, J.J. & De Vos, G.J. (1974).** Experimental tests of a basic assumption of the capture-recapture method in pond populations of Carp *Cyprinus carpio* L. *Journal of Fish Biology* 6, 317-329. **Beveridge, M.C.M., Sikdar, P.K., Frerichs, G.N. & Millar, S. (1991).** The ingestion of bacteria in suspension by the common carp *Cyprinus carpio* L. *Journal of Fish Biology* 39, 825-832. **Biagianti-Risbourg, S. (1991).** The fine structure of hepatocytes in juvenile grey mullets: *Liza saliens* Risso, *L. ramada* Risso and *L. aurata* Risso (Teleostei, Mugilidae). *Journal of Fish Biology* 39, 687-704. **Bianco, P.G. (1988).** Occurrence of the Asiatic gobionid *Pseudorasbora parva* (Temminck & Schlegel) in south-eastern Europe. *Journal of Fish Biology* 32, 973-974. **Bibby, M.C. (1972).** Population biology of the helminth parasites of *Phoxinus phoxinus* (L.), the Minnow, in a Cardiganshire lake. *Journal of Fish Biology* 4, 289-300. **Bird, D.J. & I.C. Potter, (1979a).** Metamorphosis in the paired species of lampreys, *Lampetra fluviatilis* (L.) and *Lampetra planeri* (Bloch). 1. A description of the timing and stages. *Journal of the Linnaean Society of London* 65, 145-160. **Bird, D.J. & Potter, I.C. (1979b).** Metamorphosis in the paired species of lampreys, *Lampetra fluviatilis* (L.) and *Lampetra planeri* (Bloch). 2. Quantitative data for body proportions, weights, lengths and sex ratios. *Journal of the Linnaean Society of London* 65, 127-143. **Bird, D.J., Potter, I.C., Hardisty, M.W., & Baker, B.I. (1994).** Morphology, body size and behaviour of recently metamorphosed sea lampreys, *Petromyzon marinus*, from the lower River Severn, and their relevance to the onset of parasitic feeding. *Journal of Fish Biology* 44, 67-74. **Biro, P. (1975).** The growth of bleak (*Alburnus alburnus* L.) (Pisces, Cyprinidae) in Lake Balaton and the assessment of mortality and production rate. *Annales of the Institute of Biology, Tihany* 42, 139-156. **Bodaly, R.A., Ward, R.D. & Mills, C.A. (1989).** A genetic stock study of perch, *Perca fluviatilis* L., in Windermere. *Journal of Fish Biology* 34, 965-968. **Bohlen, J. (1998).** Differences in clutch size, egg size and larval pigmentation between *Cobitis taenia* and *C. bilineata* (Cobitidae). *Italian Journal of Zoology* 65, 219-221. **Bohlen, J. (1999).** Influence of salinity on early development in the spined loach. *Journal of Fish Biology* 55, 189-198. **Boisneau, P., Mennesson, C. & Bagliniere, J.L. (1985).** Observations sur l'activite de migration de la grande alose *Alosa alosa* L. en Loire (France). *Hydrobiologia* 128, 277-284. **Boisneau, P., Mennesson-Boisneau, C. & Bagliniere, J.L. (1990).** Description d'une frayere et comportement de reproduction de la grande alose (*Alosa alosa* L.) dans le cours superieur de la Loire (France). *Bulletin Francais de Peche et Pisciculture* 316, 15-23. **Boisneau, P., Mennesson-Boisneau, C. & Guyomard, R. (1992).** Electrophoretic identity between Allis Shad, *Alosa alosa* (L.) and Twaite Shad, A. *fallax* (Lacepede). *Journal of Fish Biology* 40, 731-738. **Bourke, P., Magnan, P. & Rodriguez, M.A. (1997).** Individual variations in habitat use and morphology in brook charr. *Journal of Fish Biology* 51, 783-794. **Bracken, J.J. & Kennedy, M.P. (1967).** A key to the identification of the eggs and young stages of coarse fish in Irish waters. *Scientific Proceedings of the Royal Dublin Society* 2B, 99-108. **Brassington, R.A. & Ferguson, A. (1975).** Electrophoretic identification of Roach (*Rutilus rutilus* L.), Rudd (*Scardinius erythrophthalmus* L.), Bream (*Abramis brama* L.) and their natural hybrids. *Journal of Fish Biology* 9, 471-477. **Bregazzi, P.R. & Kennedy, C.R. (1980).** The biology of Pike, *Esox* *lucius* L., in a southern eutrophic lake. *Journal of Fish Biology* 17, 91-112. **Bregazzi, P.R. & Kennedy, C.R. (1982).** The responses of a Perch, *Perca fluviatilis* L., population to eutrophication and associated changes in fish fauna in a small lake. *Journal of Fish Biology* 20, 21-31. **Bridges, C.H. & Mullan, J.W. (1958).** A compendium of the life history and ecology of the Eastern Brook Trout *Salvelinus fontinalis* (Mitchill). *Massachussetts Division of Fish and Game Fishery Bulletin* 23, 1-30. **Broughton, N.M. & Jones, N.V. (1978).** An investigation into the growth of 0-group Roach *Rutilus rutilus* (L.) with special reference to temperature. *Journal of Fish Biology* 12, 345-358. **Brown, E.A.R. (1989).** *Growth processes in the two Scottish populations of Powan,* Coregonus lavaretus *(L.).* Unpublished PhD Thesis, University of St Andrews. **Brown, E.A.R. & Scott, D.B.C. (1987).** Abnormal pelvic fins in Scottish powan, *Coregonus lavaretus* (L.) (*Salmo*nidae, Coregoninae). *Journal of Fish Biology* 31, 443-444. **Buck, R.J.G. & Hay, D.W. (1984).** The relation between stock size and progeny of Atlantic salmon, *Salmo salar* L., in a Scottish stream. *Journal of Fish Biology* 24, 1-12. **Buck, R.J.G. & Youngson, A.F. (1982).** The downstream migration of precociously mature Atlantic Salmon, *Salmo salar* L. parr in autumn and its relation to the spawning migration of mature adult fish. *Journal of Fish Biology* 20, 279-285. **Bucke, D. (1974).** Vertebral anomalies in the common Bream *Abramis brama* (L.). *Journal of Fish Biology* 6, 681-682. **Burrough, R.J. (1978).** The population biology of two species of eyefluke, *Diplostomum spathaceum* and *Tylodelphys clavata*, in Roach and Rudd. *Journal of Fish Biology* 13, 19-32. **Burrough, R.J. & Kennedy, C.R. (1978).** Interaction of Perch (*Perca fluviatilis*) and Brown Trout (*Salmo trutta*). *Journal of Fish Biology* 13, 225-230. **Burrough, R.J. & Kennedy, C.R. (1979).** The occurrence and natural alleviation of stunting in a population of Roach, *Rutilus rutilus* (L.). *Journal of Fish Biology* 15, 93-109. **Buth, D.G. (1984).** Allozymes of the cyprinid fishes. Pages 561-590 in: Turner, B.J. (Ed.) *Evolutionary Genetics of Fishes.* Plenum, New York.

Cala, P. (1971). On the ecology of the Ide *Idus idus* (L.) in the River Kavlingean, south Sweden. *Annual Report of the Institute of Freshwater Research, Drottningholm* 50, 45-99. **Calderwood, W.L. (1930).** *Salmon and Sea Trout.* Arnold, London. **Campbell, J.S. (1977).** Spawning characteristics of Brown Trout and Sea Trout *Salmo trutta* L. in Kirk Burn, River Tweed, Scotland. *Journal of Fish Biology* 11, 217-230. **Campbell, R.D. & Branson, B.A. (1978).** Ecology and population dynamics of the black bullhead, *Ictalurus melas* (Rafinesque), in central Kentucky. *Tulane Studies in Zoology and Botany* 20, 99-136. **Campbell, R.N. (1955).** Food and feeding habits of Brown Trout, Perch and other fish in Loch Tummel. *Scottish Naturalist* 67, 23-27. **Campbell, R.N. (1957).** The effect of flooding on the growth rate of Brown Trout in Loch Tummel. *Freshwater and Salmon Fisheries Research* 14, 1-7. **Campbell, R.N. (1963).** Some effects of impoundment on the environment and growth of Brown Trout (*Salmo trutta* L.) in Loch Garry (Inverness-shire). *Freshwater and Salmon Fisheries Research* 30, 1-27. **Campbell, R.N. (1971).** The growth of brown trout, *Salmo trutta* L., in northern Scotland with special reference to the improvement of fisheries. *Journal of Fish Biology* 3, 1-28. **Campbell, R.N. (1979a).** Ferox Trout (*Salmo trutta* L.) and Charr (*Salvelinus alpinus* (L.)) in Scottish lochs. *Journal of Fish Biology* 14, 1-29.

Campbell, R.N. (1979b). Sticklebacks (*Gasterosteus aculeatus* (L.) and *Pungitius pungitius* (L.)) in the Outer Hebrides, Scotland. *Hebridean Naturalist* 3, 8-15. **Campbell, R.N. (1985).** Morphological variation in the Three-spined Stickleback (*Gasterosteus aculeatus*) in Scotland. *Behaviour* 93, 161-168. **Campbell, R.N.B. (1982).** The food of Arctic Charr in the presence and absence of Brown Trout. *Glasgow Naturalist* 20, 229-235. **Campbell, R.N.B. (1984).** Predation by Arctic Charr on the Three-spined Stickleback and its nest in Loch Meallt, Skye. *Glasgow Naturalist* 20, 409-413. **Carlander, K.D. (1977).** *Handbook of Freshwater Fishery Biology*. Iowa State University Press, Ames. **Carss, D.N., Elston, D.A., Nelson, K.C. & Kruuk, H. (1999).** Spacial and temporal trends in unexploited yellow eel stocks in two shallow lakes and associated streams. *Journal of Fish Biology* 55, 636-654. **Carss, D.N., Kruuk, H. & Conroy, J.W.H. (1990).** Predation on adult Atlantic salmon, *Salmo salar* L., by otters, *Lutra lutra* (L.), within the River Dee system, Aberdeenshire, Scotland. *Journal of Fish Biology* 37, 935-944. **Carter, C.G., Houlihan, D.F. & Owen, S.F. (1998).** Protein synthesis, nitrogen excretion and long-term growth of juvenile *Pleuronectes flesus*. *Journal of Fish Biology* 53, 272-284. **Castelnaud, G., Rochard, E., Jatteau, P. & Lepage, M. (1991).** Donnees actuelles sur la biologie d'*Acipenser sturio* dans l'estuaire de la Gironde. Pages 251-275 in: Williot, P. (Ed.) *Acipenser*. CEMAGREF, Bordeaux. **Castilho, R. & McAndrew, B.J. (1998).** Population structure of sea bass in Portugal: evidence from allozymes. *Journal of Fish Biology* 53, 1038-1049. **CEMAGREF (1997).** Restauration de l'esturgeon Europeen *Acipenser sturio*: Rapport final du programme d'execution. CEMAGREF, Bordeaux. **Chappell, L.H. (1969).** The parasites of the Three-spined Stickleback *Gasterosteus aculeatus* L. from a Yorkshire pond. I. Seasonal variation of parasite fauna. *Journal of Fish Biology* 1, 137-152. **Child, A.R., Burnell, A.M. & Wilkins, N.P. (1976).** The existence of two races of Atlantic salmon (*Salmo salar* L.) in the British Isles. *Journal of Fish Biology* 8, 35-43. **Child, A.R. & Solomon, D.J. (1977).** Observations on morphological and biochemical features of some cyprinid hybrids. *Journal of Fish Biology* 11, 125-132. **Claridge, P.N. & Gardner, D.C. (1978).** Growth and movements of the Twaite Shad, *Alosa fallax* (Lacepede), in the Severn Estuary. *Journal of Fish Biology* 12, 203-211. **Claridge, P.N., Potter, I.C. & Hughes, G.M. (1973).** Circadian rhythms of activity, ventilatory frequency and heart rate in the adult river lamprey, *Lampetra fluviatilis*. *Journal of Zoology, London* 191, 230-250. **Clelland, B. (1971).** An ecological study of a Scottish population of Bullheads. Unpublished BSc Thesis, University of Edinburgh. **Clemens, H.P. (1951a).** The food of the Burbot, *Lota lota maculosa* (LeSueur) in Lake Erie. *Transactions of the American Fisheries Society* 80, 56-66. **Clemens, H.P. (1951b).** The growth of the Burbot, *Lota lota maculosa* (LeSueur) in Lake Erie. *Transactions of the American Fisheries Society* 80, 163-173. **Clifford, S.L., McGinnity, P. & Ferguson, A. (1998).** Genetic changes in an Atlantic salmon population resulting from escaped juvenile farm salmon. *Journal of Fish Biology* 52, 118-127. **Clough, S., Garner, P., Deans, D. & Ladle, M. (1998).** Postspawning movements and habitat selection of dace in the River Frome, Dorset, southern England. *Journal of Fish Biology* 53, 1060-1070. **Cloutman, D.G. (1979).** Identification of catfish alevins of the Piedmont Carolinas. Pages 176-185 in: Wallus, R. & Voigtlander (Eds) *Proceedings of a Workshop on Freshwater Larval Fishes*. Tennessee Valley

Authority, Norris. **Coelho, M.M. (1981).** Contribution to the knowledge of the populations of *Gobio gobio* (Linnaeus, 1758) (Pisces, Cyprinidae) in Portugal. *Arquivos do Museu Bocage* 1A, 67-94. **Coles, T.F. (1981).** The distribution of Perch, *Perca fluviatilis* L., throughout their first year of life in Llyn Tegid, North Wales. *Journal of Fish Biology* 18, 15-21. **Collares-Pereira, M.J., Cowx, I.G., Sales, L.T., Pedrosa, N. & Santos-Reis, M. (1999).** Observations on the ecology of a landlocked population of allis shad in Aguieira Reservoir, Portugal. *Journal of Fish Biology* 55, 658-664. **Colle, D.E., Shireman, J.V. & Rottmann, R.W. (1978).** Food selection by Grass Carp fingerlings in a vegetated pond. *Transactions of the American Fisheries Society* 107, 149-152. **Colombo, G., Grandi, G. & Ross, R. (1984).** Gonad differentiation and body growth in *Anguilla anguilla* L. *Journal of Fish Biology* 24, 215-228. **Copp, G.H. (1990).** Shifts in the microhabitat of larval and juvenile roach, *Rutilus rutilus* (L.), in a floodplain channel. *Journal of Fish Biology* 36, 683-692. **Copp, G.H., Vaughan, C. & Wheeler, A. (1993).** First occurrence of the North American white sucker, *Catostomus commersoni* in Great Britain. *Journal of Fish Biology* 42, 615-617. **Costa, J.L., Assis, C.A., Almeida, P.R., Moreira, F.M. & Costa, M.J. (1992).** On the food of the European eel, *Anguilla anguilla* (L.), in the upper zone of the Tagus estuary, Portugal. *Journal of Fish Biology* 41, 851-858. **Courtenay, W.R. & Stauffer, J.R. (1984).** *Distribution, Biology and Management of Exotic Fishes*. John Hopkins University Press, Baltimore. **Cowx, I.G. (1983).** The biology of bream, *Abramis brama* (L.), and its natural hybrid with roach, *Rutilus rutilus* (L.), in the River Exe. *Journal of Fish Biology* 22, 631-646. **Cowx, I.G. (1988).** Distribution and variation in the growth of roach, *Rutilus rutilus* (L.), and dace, *Leuciscus leuciscus* (L.), in a river catchment in south-west England. *Journal of Fish Biology* 33, 59-72. **Cragg-Hine, D. & Jones, J.W. (1969).** The growth of Dace *Leuciscus leuciscus* (L.), Roach *Rutilus rutilus* (L.) and Chub *Squalius cephalus* (L.) in Willow Brook, Northamptonshire. *Journal of Fish Biology* 1, 59-82. **Craig, J.F. (1977).** Seasonal changes in the day and night activity of adult Perch, *Perca fluviatilis* L. *Journal of Fish Biology* 11, 161-166. **Craig, J.F. (1987).** *The Biology of Perch and Related Fish*. Croom Helm, Beckenham. **Craig, J.F. & Kipling, C. (1983).** Reproduction effort versus the environment: case histories of Windermere Perch, *Perca fluviatilis* L., and Pike, *Esox lucius* L. *Journal of Fish Biology* 22, 713-727. **Crisp, D.T. (1963).** A preliminary survey of Brown Trout (*Salmo trutta* L.) and Bullheads (*Cottus gobio* L.) in high altitude becks. *Salmon and Trout Magazine* 167, 45-49. **Crisp, D.T., Mann, R.H.K. & McCormack, J.C. (1975).** The populations of fish in the River Tees system on the Moor House National Nature Reserve, Westmorland. *Journal of Fish Biology* 7, 573-594. **Crivelli, A.J. (1981).** The biology of the Common Carp, *Cyprinus carpio* L., in the Camargue, southern France. *Journal of Fish Biology* 18, 271-290. **Cross, D.G. (1969).** Aquatic weed control using Grass Carp. *Journal of Fish Biology* 1, 27-30. **Cross, D.G. (1970).** The tolerance of Grass Carp *Ctenopharyngodon idella* (Val.) to seawater. *Journal of Fish Biology* 2, 231-233. **Crozier, W.W. & Ferguson, A. (1986).** Electrophoretic examination of the population structure of Brown Trout, *Salmo trutta* L., from the Lough Neagh catchment, Northern Ireland. *Journal of Fish Biology* 28, 459-478. **Cunjak, R.A. & Power, G. (1986).** Seasonal changes in the physiology of brook trout, *Salvelinus fontinalis* (Mitchill), in a sub-Arctic river system. *Journal of Fish Biology* 29, 278-288.

Dabrowski, K.R. & Jewson, D.H. (1984). The influence of light environment on depth of visual feeding by larvae and fry of *Coregonus pollan* (Thompson) in Lough Neagh. *Journal of Fish Biology* 25, 173-182. **Dabrowski, K.R., Kaushik, S.J. & Luquet, P. (1984).** Metabolic utilisation of body stores during the early life of Whitefish, *Coregonus lavaretus* L. *Journal of Fish Biology* 23, 721-730. **Dando, P.R. & Demir, N. (1985).** On the spawning and nursery grounds of Bass, *Dicentrarchus labrax*, in the Plymouth area. *Journal of the Marine Biological Association of the UK* 65, 159-168. **Danzmann, R.G., Ihssen, P.E. & Herbert, P.D.N. (1991).** Genetic discrimination of wild and hatchery populations of brook charr, *Salvelinus fontinalis* (Mitchill), in Ontario using mitochondrial DNA analysis. *Journal of Fish Biology* 39A, 69-78. **Darchambeau, F. & Poncia, P. (1997).** Field observations of the spawning behaviour of European grayling. *Journal of Fish Biology* 51, 1066-1068. **Dartnell, H.J.G. (1973).** Parasites of the Nine-spined Stickleback *Pungitius pungitius* (L.). *Journal of Fish Biology* 5, 505-510. **Davies, C.E., Shelley, J., Harding, P.T., McLean, I.F.G., Gardiner, R. & Peirson, G. (2004).** *Freshwater Fishes in Britain.* Harley, London. **Davies, P.M.C. (1963).** Food input and energy extraction efficiency in *Carassius auratus*. *Nature, London* 198, 707. **De Groot, S.J. (1989).** Decline of the catches of coregonids and migratory smelt in the lower Rhine, the Netherlands. *ICES Anadromous and Catadromous Fish Committee Publication CM* 1989/M:18, 1-11. **De Groot, S.J. (1990).** The former Allis and Twaite Shad fisheries of the lower Rhine, the Netherlands. *Journal of Applied Ichthyology* 6, 252-256. **Department of the Environment (1995).** *The Habitats Directive: How it will apply to Great Britain.* Department of the Environment, London. **De Silva, S.S. (1980).** Biology of juvenile grey mullet: a short review. *Aquaculture* 19, 21-36. **De Wit, J.J.D. (1954).** The story of Bitterling and mussel. *The Aquarist* July, 1954, 75-77. **Dembinski, W. (1971).** Vertical distribution of Vendace *Coregonus albula* L. and other pelagic fish species in some Polish lakes. *Journal of Fish Biology* 3, 341-357. **Desaunau, Y. & Guerault, D. (1997).** Seasonal and long-term changes in biometrics of eel larvae: a possible relationship between recruitment variation and North Atlantic ecosystem productivity. *Journal of Fish Biology* 51A, 317-339. **Diana, J.S. (1984).** The growth of Largemouth Bass, *Micropterus salmoides* (Lacepede), under constant and fluctuating temperatures. *Journal of Fish Biology* 24, 165-172. **Dodge, D.P. & MacCrimmon, H.R. (1970).** Vital statistics of a population of Great Lakes rainbow trout (*Salmo gairdneri*) characterised by an extended spawning season. *Journal of the Fisheries Research Board of Canada* 27, 613-618. **Drost, M.R. & Van den Boogaart, J.G.M. (1986).** The energetics of feeding strikes in larval carp, *Cyprinus carpio*. *Journal of Fish Biology* 29, 371-380.

Ede, S & Carlson, C.A. (1977). Food habits of Carp and White Sucker in South Platte and St Vrain Rivers and Goosquill Pond, Weld County, Colorado. *Transactions of the American Fisheries Society* 106, 339-346. **Edwards, D.J. (1973).** Aquarium studies on the consumption of small animals by 0-group Grass Carp, *Ctenopharyngodon idella* (Val.). *Journal of Fish Biology* 5, 599-606. **Egglishaw, H.J. (1967).** The food, growth and population structure of Salmon and Trout in two streams in the Scottish highlands. *Freshwater and Salmon Fisheries Research* 38, 1-32. **Eiras, J.C. (1981).** Sur une population d'*Alosa alosa* L., poisson migrateur amphibiotique, thalassotrophe, bloquee en eau douce au Portugal. *Cybium* 5, 69-73. **Eiras, J.C. (1983).** Some aspects on the biology of a landlocked population of anadromous shad: *Alosa alosa* L. *Publicaciones Istituto Zoologia Faculdade Ciencias do Porto* 180, 1-16. **Eklov, A.G., Greenberg, L.A., Larsson, P. & Berglund, O. (1999).** Influence of water quality, habitat and species richness on brown trout populations. *Journal of Fish Biology* 54, 33-43. **Elliott, J.M. (1976).** The downstream drifting of eggs of brown trout, *Salmo trutta* L. *Journal of Fish Biology* 9, 45-50. **Elliott, J.M. (1994).** *Quantitative Ecology and the Brown Trout.* Oxford University Press, Oxford. **Ellison, F.B. (1935).** Shad. *Transactions of the Woolhope Naturalists Field Club* 1935, 135-139. **Ellison, N.F. (1966).** Notes on Lakeland Schelly. *Changing Scene* 3, 46-53. **Ellison, N.F. & Chubb, J.C. (1968).** The Smelt of Rostherne Mere, Cheshire. *Lancashire and Cheshire Fauna Society* 53, 7-16. **Ellison, N.F. & Cooper, J.R. (1967).** Further notes on Lakeland Schelly. *Field Naturalist* 12. **Enderlein, O. (1989).** Migratory behaviour of adult cisco, *Coregonus albula* L., in the Bothnian Bay. *Journal of Fish Biology* 34, 11-18. **Eneqvist, P. (1938).** The Brook Lamprey as an ecological modification of the River Lamprey. On the River and Brook Lampreys of Sweden. *Arkiv for Zoologi* 29A, 1-22. **Environmental Protection Agency (2001).** *Register of Hydrometric Gauging Stations in Ireland.* Environmental Protection Agency, Dublin. **Erman, F. (1961).** On the biology of the Thick-lipped Mullet (*Mugil chelo*). *Rapport et Proces-vebaux des Reunions Conseil Permanent International pour l'Exploration Scientifique de la Mer* 16, 277-285. **Etnier, D.A. (1971).** Food of three species of sunfishes (*Lepomis*, Centrarchidae) and their hybrids in three Minnesota lakes. *Transactions of the American Fisheries Society* 100, 124-128.

Fabricius, E. & Gustavson, K.J. (1955). Observations on the spawning behaviour of the Grayling, *Thymallus thymallus* (L.). *Annual Report of the Institute of Freshwater Research, Drottningholm* 36, 75-103. **Fabricius, E. & Gustavson, K.J. (1958).** Some new observations on the spawning behaviour of the Pike, *Esox lucius* L. *Annual Report of the Institute of Freshwater Research, Drottningholm* 39, 23-54. **Fahy, E. (1977).** Characteristics of the freshwater occurrence of sea trout *Salmo trutta* in Ireland. *Journal of Fish Biology* 11, 635-646. **Fahy, E. (1978).** Variations in some biological characteristics of British sea trout, *Salmo trutta* L. *Journal of Fish Biology* 13, 123-138. **Farr-Cox, F., Leonard, S. & Wheeler, A. (1996).** The status of the recently introduced fish *Leucaspius delineatus* (Cyprinidae) in Great Britain. *Fisheries Management and Ecology* 3, 193-199. **Farrugio, H. (1977).** Annotated key for determination of Mugilidae adults and alevins of Tunisia. *Cybium* 3, 57-74. **Fechhelm, R.G. Fitzgerald, P.S., Bryan, J.D. & Gallaway, B.J. (1993).** Effect of salinity and temperature on the growth of yearling Arctic cisco (*Coregonus autumnalis*) of the Alaskan Beaufort Sea. *Journal of Fish Biology* 43, 463-474. **Fedorova, G.V. & Vetkasov, S.A. (1973).** The biological characteristics and abundance of the Lake Ilmen Ruffe, *Acerina cernua*. *Journal of Ichthyology* 14, 836-841. **Ferguson, A. (1974).** The genetic relationships of the Coregonid fishes of Britain and Ireland indicated by electrophoretic analysis of tissue proteins. *Journal of Fish Biology* 6, 311-315. **Ferguson, A., Himberg, K.J.M. & Svardson, G. (1978).** The systematics of the Irish Pollan (*Coregonus pollan* Thompson): an electrophoretic comparison with other Holarctic Coregoninae. *Journal of Fish Biology* 12, 221-233. **Ferguson, A. & Mason, F.M. (1981).** Allozyme evidence for reproductively isolated sympatric populations of Brown Trout *Salmo trutta* L. in Lough Melvin, Ireland. *Journal of Fish Biology* 18, 629-642. **Fickling, N.J. (1982).** The identification of Pike by means of characteristic marks. *Fisheries Management* 13, 79-82. **Fickling, N.J. & Lee, R.L.G. (1985).** A study of the movements of the Zander, *Lucioperca lucioperca* L., population of two lowland fisheries. *Aquaculture and Fisheries Management* 16, 377-393. **Fitzmaurice, P. (1983a).** Carp (*Cyprinus carpio* L.) in Ireland. *Irish Fisheries Investigations* 23A, 5-10. **Fitzmaurice, P. (1983b).** Some aspects of the biology and management of Pike (*Esox lucius* L.) stocks in Irish fisheries. *Journal of Life Sciences, Royal Dublin Society* **4, 161-173**. **Flowerdew, M.W. & Grove, D.J. (1980).** An energy budget for juvenile Thick-lipped Mullet, *Crenimugil labrosus* (Risso). *Journal of Fish Biology* 17, 395-410. **Fouda, M.M. (1979).** Studies on scale structure in the Common Goby *Pomatoschistus microps* Kroyer. *Journal of Fish Biology* 15, 165-172. **Fouda, M.M. & Miller, P.J. (1979).** Alkaline phosphatase activity in the skin of the Common Goby, *Pomatoschistus microps*, in relation to cycles in scale and body growth. *Journal of Fish Biology* 15, 263-274. **Fox, P.J. (1978).** Preliminary observations on different reproduction strategies in the Bullhead (*Cottus gobio* L.) in northern and southern England. *Journal of Fish Biology* 12, 5-11. **Friend, G.F. (1956).** A new subspecies of char from Loch Eck. *Glasgow Naturalist* 17, 219-220. **Frost, W.E. (1939).** River Liffey Survey. II. The food consumed by the Brown Trout (*Salmo trutta* L.) in acid and alkaline water. *Proceedings of the Royal Irish Academy* 45, 139-206. **Frost, W.E. (1940).** Rainbows of a peat lough on Arranmore. *Salmon and Trout Magazine* 100, 234-240. **Frost, W.E. (1943).** The natural history of the Minnow *Phoxinus phoxinus*. *Journal of Animal Ecology* 12, 139-162. **Frost, W.E. (1945).** The age and growth of Eels (*Anguilla anguilla*) from the Windermere catchment area. *Journal of Animal Ecology* 14, 26-36, 106-124. **Frost, W.E. (1974).** *A Survey of the Rainbow Trout (*Salmo gairdneri*) in Britain and Ireland.* Salmon & Trout Association, London. **Frost, W.E. (1977).** The food of Charr, *Salvelinus willughbii* (Gunther), in Windermere. *Journal of Fish Biology* 11, 531-548. **Frost, W.E. & Brown, M.E. (1967).** *The Trout*. Collins, London. **Frost, W.E. & Kipling, C. (1980).** The growth of Charr, *Salvelinus willughbii*, Gunther, in Windermere. *Journal of Fish Biology* 16, 279-290. **Fuller, J.D. & Scott, D.B.C. (1976).** The reproductive cycle of *Coregonus lavaretus* (L.) in Loch Lomond, Scotland, in relation to seasonal changes in plasma cortisol concentration. *Journal of Fish Biology* 9, 105-117.

Galloway, M.L. & Kilambi, R.V. (1988). Thermal enrichment of a reservoir and the effects on annulus formation and growth of largemouth bass, *Micropterus salmoides*. *Journal of Fish Biology* 32, 533-544. **Garcia, D.L.C. & Verspoor, E. (1989).** Natural hybridization between Atlantic salmon, *Salmo salar* L., and brown trout, *Salmo trutta* L., in northern Spain. *Journal of Fish Biology* 34, 41-46. **Garcia, L.M. & Martin, F.D. (1985).** An in situ estimate of daily food consumption and alimentary canal evacuation rates of common carp, *Cyprinus carpio* L. *Journal of Fish Biology* 27, 487-494. **Gardiner, W.R. (1974).** An electrophoretic method for distinguishing the young fry of salmon *Salmo salar* (L.) from those of trout *Salmo trutta* (L.). *Journal of Fish Biology* 6, 517-519. **Gardner, A.S., Walker, A.F. & Greer, R.B. (1988).** Morphometric analysis of two ecologically distinct forms of Arctic Charr, *Salvelinus alpinus*

(L.), in Loch Rannoch, Scotland. *Journal of Fish Biology* 32, 901-910. **Garnas, E. (1982).** Growth of different year classes of Smelt *Osmerus eperlanus* L. in Lake Tyrifjorden, Norway. *Fauna Norvegica* 3, 1-6. **Garner, P. Clough, S., Griffiths, S.W., Deans, D. & Ibbotson, A. (1998).** Use of shallow marginal habitat by *Phoxinus phoxinus*: a trade-off between temperature and food? *Journal of Fish Biology* 52, 600-609. **Gaudreault, A., Miller, T., Montgomery, L. & Fitzgerald, G.J. (1986).** Interspecific interactions and diet of sympatric juvenile brook charr, *Salvelinus fontinalis*, and adult ninespine sticklebacks, *Pungitius pungitius*. *Journal of Fish Biology* 28, 133-140. **Gee, A.S., Milner, N.J. & Hemsworth, R.J. (1978).** The production of juvenile salmon, *Salmo salar* in the upper Wye, Wales. *Journal of Fish Biology* 13, 439-451. **George, D.G. & Winfield, I.J. (2000).** Factors influencing the spatial distribution of zooplankton and fish in Loch Ness, UK. *Freshwater Biology* 44, 557-570. **George, E.L. & Hadley, W.F. (1979).** Food and habitat partitioning between Rock Bass (*Ambloplites rupestris*) and Smallmouth Bass (*Micropterus dolomieu*) young of the year. *Transactions of the American Fisheries Society* 108, 253-261. **Gerlach, J.M. (1983).** Characters for distinguishing larvae of carp, *Cyprinus carpio*, and goldfish, *Carassius auratus*. *Copeia* 1983, 116-121. **Gerrish, C.S. (1936–39).** Scales of Avon Trout and Grayling. *Report of Avon Biological Research* 3, 81-95; 4, 44-58; 5, 70-78; 6, 54-59. **Gervers, F.W.K. (1954).** A supernumerary pelvic fin in the powan (*Coregonus clupeoides* Lacepede). *Nature, London* 174, 935. **Ghan, D. & Sprules, W.G. (1993).** Diet, prey selection and growth of larval and juvenile burbot *Lota lota* (L.). *Journal of Fish Biology* 42, 47-64. **Gibson, D.L. (1972).** Flounder parasites as biological tags. *Journal of Fish Biology* 4, 1-10. **Giles, N. (1983).** The possible role of environmental calcium levels during the evolution of phenotypic diversity in Outer Hebridean populations of Three-spined Sticklebacks, *Gasterosteus aculeatus*. *Journal of Zoology, London* 199, 535-545. **Giles, N., Street, M. & Wright, R.M. (1990).** Diet composition and prey preference of tench, *Tinca tinca* (L.), common bream, *Abramis brama* (L.), perch, *Perca fluviatilis* L. and roach *Rutilus rutilus* (L.) in two contrasting gravel pit lakes: potential trophic overlap with wildfowl. *Journal of Fish Biology* 37, 945-958. **Goldspink, C.R. (1978).** Comparative observations on the growth rate and year class strength of Roach *Rutilus rutilus* L. in two Cheshire lakes, England. *Journal of Fish Biology* 12, 421-433. **Goldspink, C.R. (1981).** A note on the growth-rate and year-class strength of Bream, *Abramis brama* L., in three eutrophic lakes, England. *Journal of Fish Biology* 19, 665-674. **Goldspink, C.R. (1990).** The distribution and abundance of young (I+–II+) perch *Perca fluviatilis* L., in a deep eutrophic lake, England. *Journal of Fish Biology* 36, 439-448. **Goldspink, C.R. & Goodwin, D.A. (1979).** A note on age composition, growth rate and food of Perch, *Perca fluviatilis* (L.), in four eutrophic lakes, England. *Journal of Fish Biology* 14, 489-505. **Gozlan, R.E., Pinder, A.C. & Shelly, J. (2002).** Occurrence of the Asiatic cyprinid *Pseudorasbora parva* in England. *Journal of Fish Biology* 61, 298-300. **Graham, J.H. & Hastings, R.W. (1984).** Distributional patterns of sunfishes on the New Jersey Coastal Plain. *Environmental Biology of Fishes* 10, 137-148. **Graham, T.T. & Jones, J.W. (1962).** The biology of Llyn Tegid Trout 1960. *Proceedings of the Zoological Society of London* 139, 657-683. **Greenbank, J. & Nelson, P. (1959).** Life history of the Three-spine Stickleback *Gasterosteus aculeatus* Linnaeus in

Karluk Lake and Bare Lake Kodiak Island, Alaska. *U.S. Fish and Wildlife Service Bulletin* 153, 537-559. **Greenwood, M.F.D. & Metcalfe, N.B. (1998).** Minnows become nocturnal at low temperatures. *Journal of Fish Biology* 53, 25-32. **Griffiths, S.W. (1997).** Preferences for familiar fish do not change with predation risk in the European minnow. *Journal of Fish Biology* 51, 489-495. **Griswold, B.L. & Smith, L.L. (1973).** The life history and trophic relationship of the Nine-spine Stickleback, *Pungitius pungitius*, in the Apostle Islands area of Lake Superior. *Fisheries Bulletin* 71, 1039-1060. **Guinea, J. & Fernandez, F. (1992).** Morphological and biometrical study of the gill rakers in four species of mullet. *Journal of Fish Biology* 41, 381-398.

Hagelin, L.O., (1959). Further aquarium observations on the spawning habits of the River Lamprey (*Petromyzon fluviatilis*). *Oikos* 10: 50-64. **Hamilton, K.E., Ferguson, A., Taggart, J.B., Tomasson, T., Walker, A. & Fahy, E. (1989).** Post-glacial colonisation of brown trout, *Salmo trutta* L.: *Ldh-5* as a phylogeographic marker locus. *Journal of Fish Biology* 35, 651-664. **Hancock, R.S., Jones, J.W. & Shaw R. (1976).** A preliminary report on the spawning behaviour and nature of sexual selection in the Barbel, *Barbus barbus* (L.). *Journal of Fish Biology* 9, 21-28. **Hansen, L.P. & Pethon, P. (1985).** The food of Atlantic salmon, *Salmo salar* L., caught by long-line in northern Norwegian waters. *Journal of Fish Biology* 26, 553-562. **Haram, O.J. (1968).** *A preliminary investigation of the biology of the Gwyniad (*Coregonus *sp.) of Llyn Tegid.* Unpublished PhD Thesis, University of Liverpool. **Haram, O.J. & Jones, J.W. (1971).** Some observations on the food of the Gwyniad, *Coregonus pennantii* Valenciennes of Llyn Tegid (Lake Bala), North Wales. *Journal of Fish Biology* 3, 287-295. **Hardie, R.P. (1940).** *Ferox and Char in the Lochs of Scotland.* Oliver and Boyd, Edinburgh. **Hardisty, M.W. (1944).** The life-history and growth of the Brook Lamprey (*Lampetra planeri*). *Journal of Animal Ecology* 13, 110-122. **Hardisty, M.W. (1961a).** The growth of larval lampreys. *Journal of Animal Ecology* 30, 357-371. **Hardisty, M.W. (1961b).** Studies on an isolated spawning population of the brook lamprey (*Lampetra planeri*). *Journal of Animal Ecology* 30, 339-355. **Hardisty, M.W. (1961c).** Oocyte number as a diagnostic character for the identification of ammocoete species. *Nature, London* 191, 1215-1216. **Hardisty, M.W. (1964).** The fecundity of lampreys. *Archiv für Hydrobiologie* 60, 340-357. **Hardisty, M.W. (1969).** Information on the growth of the ammocoete larvae of the anadromous Sea Lamprey *Petromyzon marinus* in British rivers. *Journal of Zoology, London* 159: 139-144. **Hardisty, M.W. (1970).** The relationship of gonadal development to the life cycles of the paired species of lamprey, *Lampetra fluviatilis* (L.) and *Lampetra planeri* (Bloch). *Journal of Fish Biology* 2, 173-181. **Hardisty, M.W. & Huggins, R.L. (1970).** Larval growth in the river lamprey, *Lampetra fluviatilis*. *Journal of Zoology, London* 161, 549-559. **Hardisty, M.W. & Potter, I.C. (Eds) (1971).** *The Biology of Lampreys*. Academic Press, London. **Hardisty, M.W., Potter, I.C. & Sturge, R. (1970).** A comparison of the metamorphosing and macrophthalmia stages in the lampreys *Lampetra fluviatilis* and *L. planeri*. *Journal of Zoology, London* 162, 383-400. **Hardy, E. (1954).** The bitterling in Lancashire. *Salmon and Trout Magazine* 142, 548-553. **Hardy, J.D. (1978).** Development of fishes of the Mid-Atlantic Bight. An atlas of egg, larval and juvenile stages. *U.S. Fish and Wildlife Service Biological Services Program* FWS-OBS-78/12.

Harris, M.T. & Wheeler, A. (1974). *Ligula* infestation of Bleak *Alburnus alburnus* (L.) in the tidal Thames. *Journal of Fish Biology* 6, 181-188. **Harrod, C., Griffiths, D., McCarthy, T.K. & Rosell, R. (2001).** The Irish pollan, *Coregonus autumnalis*: options for its conservation. *Journal of Fish Biology* 59A, 339-355. **Hartley, P.H.T. (1947a).** The natural history of some British freshwater fishes. *Proceedings of the Zoological Society of London* 117, 129-206. **Hartley, P.H.T. (1947b).** The coarse fishes of Britain. *Scientific Publications of the Freshwater Biological Association* 12, 1-40. **Hartley, P.H.T. (1948).** Food and feeding relationships in a community of freshwater fishes. *Journal of Animal Ecology* 17, 1-14. **Hartley, S.E., Bartlett, S.E. & Davidson, W.S. (1992a).** Mitochondrial DNA analysis of Scottish populations of Arctic charr, *Salvelinus alpinus* (L.). *Journal of Fish Biology* 40, 219-224. **Hartley, S.E., McGowan, C., Greer, R.B. & Walker, A.F. (1992b).** The genetics of sympatric Arctic charr (*Salvelinus alpinus* (L.)) populations from Loch Rannoch, Scotland. *Journal of Fish Biology* 41, 1021-1032. **Hazen, T.C. & Esch, G.W. (1978).** Observations on the ecology of *Clinostomum marginatum* in Largemouth Bass (*Micropterus salmoides*). *Journal of Fish Biology* 12, 411-420. **Healey, M.C. (1972).** On the population ecology of the Common Goby in the Ythan Estuary. *Journal of Natural History* 6, 133-145. **Healy, A. (1956).** Pike (*Esox lucius* L.) in three Irish Lakes. *Scientific Proceedings of the Royal Dublin Society* 27, 51-63. **Healy, J.A. & Mulcahy, M.F. (1980).** A biochemical genetic analysis of populations of the Northern Pike, *Esox lucius* L., from Europe and North America. *Journal of Fish Biology* 17, 317-324. **Heap, S.P. & Goldspink, G. (1986).** Alterations to the swimming performance of carp, *Cyprinus carpio*, as a result of temperature acclimation. *Journal of Fish Biology* 29, 747-754. **Hellawell, J.M. (1969).** Age determination and growth of the Grayling *Thymallus thymallus* (L.) of the River Lugg, Herefordshire. *Journal of Fish Biology* 1, 373-382. **Hellawell, J.M. (1971a).** The food of Grayling *Thymallus thymallus* (L.) of the River Lugg, Herefordshire. *Journal of Fish Biology* 3, 187-197. **Hellawell, J.M. (1971b).** The autecology of the chub, *Squalius cephalus* (L.), of the River Lugg and Afon Llynfi. I. Age determination, population structure and growth. *Freshwater Biology* 1, 29-60. **Hellawell, J.M. (1971c).** The autecology of the chub, *Squalius cephalus* (L.), of the River Lugg and Afon Llynfi. II. Reproduction. *Freshwater Biology* 1, 135-148. **Hellawell, J.M. (1971d).** The autecology of the chub, *Squalius cephalus* (L.), of the River Lugg and Afon Llynfi. III. Diet and feeding habits. *Freshwater Biology* 1, 369-387. **Hellawell, J.M. (1972).** The growth, reproduction and food of the Roach *Rutilus rutilus* (L.), of the River Lugg, Herefordshire. *Journal of Fish Biology* 4, 469-486. **Hellawell, J.M. (1974).** The ecology of populations of Dace, *Leuciscus leuciscus* (L.), from two tributaries of the River Wye, Herefordshire, England. *Freshwater Biology* 4, 577-604. **Hervey, G.F. & Hems, J. (1968).** *The Goldfish.* Fabre and Fabre, London. **Hesthagen, T. & Jonsson, B. (1998).** The relative abundance of brown trout in acidic softwater lakes in relation to water quality in tributary streams. *Journal of Fish Biology* 52, 419-429. **Hickley, P. & Bailey, R.G. (1982).** Observations on the growth and production of Chub *Leuciscus cephalus* and Dace *Leuciscus leuciscus* in a small lowland river in southeast England, UK. *Freshwater Biology* 12, 167-178. **Hickling, C.F. (1970).** A contribution to the natural history of the English Grey Mullets, Pisces Mugilidae. *Journal of the Marine Biological*

Association of the UK 50, 609-633. **Hile, R. (1941).** Age and growth of the Rock Bass, *Ambloplites rupestris* (Rafinesque), in Nebish Lake, Wisconsin. *Transactions of the Wisconsin Academy of Science, Arts and Letters* 33, 189-337. **Hindar, K. & Nordland, J. (1989).** A female Atlantic salmon, *Salmo salar* L., maturing sexually in the parr stage. *Journal of Fish Biology* 35, 461-464. **Hinkens, E. & Cochrane, P.A. (1988).** Taste buds on pelvic ray fins of the Burbot, *Lota lota* (L.). *Journal of Fish Biology* 32, 975. **Holden, M.J. & Williams, T. (1974).** The biology, movements and population dynamics of Bass *Dicentrarchus labrax* in English waters. *Journal of the Marine Biological Association of the UK* 54, 91-107. **Houghton, W. (1879).** *British Freshwater Fishes*. Mackenzie, London. **Huggins, R.J. & Thompson, A. (1970).** Communal spawning of brook and river lampreys, *Lampetra planeri* Bloch and *Lampetra fluviatilis* L. *Journal of Fish Biology* 2, 53-54. **Hunt, P.C. & Jones, J.W. (1974a).** A population study of Barbel *Barbus barbus* (L.) in the River Severn, England. I. Densities. *Journal of Fish Biology* 6, 255-267. **Hunt, P.C. & Jones, J.W. (1974b).** A population study of Barbel *Barbus barbus* (L.) in the River Severn, England. II. Movements. *Journal of Fish Biology* 6, 269-278. **Hunt, P.C. & Jones, J.W. (1975).** A population study of *Barbus barbus* L. in the River Severn, England. III. Growth. *Journal of Fish Biology* 7, 361-376. **Hunt, P.C. & O'Hara, K. (1973).** Overwinter feeding in Rainbow Trout. *Journal of Fish Biology* 5, 277-280. **Hurnell, R.H. & Price, D.J. (1991).** Natural hybrids between Atlantic salmon, *Salmo salar* L., and brown trout, *Salmo trutta* L., in juvenile salmonid populations in south-west England. *Journal of Fish Biology* 39A, 335-342. **Hutchinson, P. (1983).** Some ecological aspects of the Smelt, *Osmerus eperlanus* (L.), from the River Cree, southwest Scotland. *Proceedings of the British Coarse Fish Conference, Liverpool* 3, 178-191. **Hutchinson, P. & Mills, D.H. (1987).** Characteristics of spawning-run smelt, *Osmerus eperlanus* (L.), from a Scottish river, with recommendations for their conservation and management. *Aquaculture and Fisheries Management* 18, 249-258. **Hutchinson, S. & Hawkins, L.E. (1993).** The migration and growth of 0-group flounders *Platichthys flesus* in mixohaline conditions. *Journal of Fish Biology* 43, 325-328. **Hutton, J.A. (1923).** Something about Grayling scales. *Salmon and Trout Magazine* January, 3-8. **Hynd, I.J.R. (1964).** Large sea trout from the Tweed district. *Salmon and Trout Magazine* 172, 151-154. **Hynes, H.B.N. (1950).** The food of freshwater sticklebacks (*Gasterosteus aculeatus* and *Pungitius pungitius*), with a review of methods used in studies of the food of fishes. *Journal of Animal Ecology* 19, 36-58. **Hyslop, E.J. (1982).** The feeding habits of 0+ Stone Loach, *Noemacheilus barbatulus* (L.) and Bullhead, *Cottus gobio* L. *Journal of Fish Biology* 21, 187-196.

Ibrahim, A.A. & Huntingford, F.A. (1988). Foraging efficiency in relation to within-species variation in morphology in three-spined sticklebacks, *Gasterosteus aculeatus*. *Journal of Fish Biology* 33, 823-824. **IUCN (1994).** *IUCN Red List Categories*. International Union for the Conservation of Nature and Natural Resources, Gland. **IUCN (2001).** *Global Strategy on Invasive Alien Species*. International Union for the Conservation of Nature and Natural Resources, Gland. **Ivanova-Berg, M.M. (1962).** On lamprey larvae. *Voprosy Ikhtiologii* 2, 557.

Jackman, L.A.J. (1954). The early development stages of the Bass *Morone labrax. Proceedings of the Zoological Society of London* 124, 531-534. **Jenkins, J.T. (1925).** *The Fishes of the British Isles, both Fresh and Salt.* Warne, London. **Jenkins, R.E. & Burkhead, N.M. (1993).** *Freshwater Fishes of Virginia.* American Fisheries Society, Bethesda. **Johansson, L. (1987).** Experimental evidence for interactive habitat segregation between roach (*Rutilus rutilus*) and rudd (*Scardinius erythrophthalmus*) in a shallow eutrophic lake. *Oecologia* 73, 21-27. **Jones, J.W. (1953).** Part I. Scales of Roach. Part II. Age and growth of the Trout, Grayling, Perch and Roach of Llyn Tegid (Bala) and the Roach from the River Birket. *Fisheries Investigations, London* 5, 1-8. **Jones, J.W. (1959).** *The Salmon.* Collins, London. **Jones, J.W. & Hynes, H.B.N. (1950).** The age and growth of *Gasterosteus aculeatus*, *Pungitius pungitius* and *Spinachia vulgaris*, as shown by their otoliths. *Journal of Animal Ecology* 19, 59-73. **Jones, N.S. (1952).** The bottom fauna and the food of flatfish off the Cumberland coast. *Journal of Animal Ecology* 21, 182-205. **Jurvelius, J., Lindem, T. & Heikkinen, T. (1988).** The size of a Vendace, *Coregonus albula* L., stock in a deep lake basin monitored by hydro-acoustic methods. *Journal of Fish Biology* 32, 679-687.

Kainua, K & Valtonen, T. (1980). Distribution and abundance of European River Lamprey (*Lampetra fluviatilis*) larvae in three rivers running into Bothnian Bay, Finland. *Canadian Journal of Fisheries and Aquatic Sciences* 37, 1960-1966. **Karas, P. (1990).** Seasonal changes in growth and standard metabolic rate of juvenile perch, *Perca fluviatilis* L. *Journal of Fish Biology* 37, 913-920. **Keast, A. & Eadie, J.M. (1985).** Growth depensation in year-0 Largemouth Bass: the influence of diet. *Transactions of the American Fisheries Society* 114, 204-213. **Keast, A. & Welsh, L. (1968).** Daily feeding periodicities, food uptake rates and dietary changes with hour of day in some lake fishes. *Journal of the Fisheries Research Board of Canada* 25, 1133-1144. **Kelley, D. (1979).** Bass populations and movements on the west coast of the U.K. *Journal of the Marine Biological Association of the UK* 59, 889-936. **Kelley, D. (1986).** Bass nurseries on the west coast of the U.K. *Journal of the Marine Biological Association of the UK* 66, 439-464. **Kennedy, C.R. (1969).** Tubificid oligochaetes as food of Dace *Leuciscus leuciscus* (L.). *Journal of Fish Biology* 1, 11-15. **Kennedy, C.R. (1981).** The occurrence of *Eubothrium fragile* (Cestoida: Pseudophyllidae) in Twaite Shad, *Alosa fallax* (Lacepede) in the River Severn. *Journal of Fish Biology* 19, 171-178. **Kennedy, C.R. (1984a).** The status of flounders, *Platichthys flesus* L., as hosts of the acanthocephalan *Pomphorhynchus laevis* (Muller), and its survival in marine conditions. *Journal of Fish Biology* 24, 135-150. **Kennedy, C.R. (1984b).** The dynamics of a declining population of the acanthocephalan *Acanthocephalus clavula* in Eels, *Anguilla anguilla*, in a small river. *Journal of Fish Biology* 25, 665-677. **Kennedy, C.R. & Hine, P.M. (1969).** Population biology of the cestode *Proteocephalus torulosus* (Batsch) in dace *Leuciscus leuciscus* (L.) of the River Avon. *Journal of Fish Biology* 1, 209-219. **Kennedy, G.J.A. & Strange, C.D. (1978).** Seven years on – a continuing investigation of salmonid stocks in Lough Erne tributaries. *Journal of Fish Biology* 12, 325-330. **Kennedy, M. (1969).** Spawning and early development of the Dace *Leuciscus leuciscus* (L.). *Journal of Fish Biology* 1, 249-259. **Kennedy, M. & Fitzmaurice, P. (1968a).** The biology of the Bream *Abramis* *abrama* (L.) in Irish waters. *Proceedings of the Royal Irish Academy* 67B, 95-161. **Kennedy, M. & Fitzmaurice, P. (1968b).** Occurrence of eggs of Bass *Dicentrarchus labrax* on the southern coasts of Ireland. *Journal of the Marine Biological Association of the UK* 48, 585-592. **Kennedy, M. & Fitzmaurice, P. (1969).** Age and growth of Thick-lipped Grey Mullet *Crenimugil labrosus* in Irish waters. *Journal of the Marine Biological Association of the UK* 1, 683-699. **Kennedy, M. & Fitzmaurice, P. (1970).** The biology of Tench *Tinca tinca* (L.) in Irish waters. *Proceedings of the Royal Irish Academy* 69B, 31-64. **Kennedy, M. & Fitzmaurice, P. (1972a).** Some aspects of the biology of Gudgeon *Gobio gobio* (L.) in Irish waters. *Journal of Fish Biology* 4, 425-440. **Kennedy, M. & Fitzmaurice, P. (1972b).** The biology of the Bass, *Dicentrarchus labrax*, in Irish waters. *Journal of the Marine Biological Association of the UK* 52, 557-597. **Kennedy, M. & Fitzmaurice, P. (1974).** Biology of Rudd *Scardinius erythrophthalmus* (L.) in Irish waters. *Proceedings of the Royal Irish Academy* 74B, 246-282. **Kilambi, R.V. & Robison, W.R. (1979).** Effects of temperature and stocking density on food consumption and growth of Grass Carp *Ctenopharyngodon idella*, Val. *Journal of Fish Biology* 15, 337-342. **Kipling, C. (1984).** Some observations on autumn-spawning charr, *Salvelinus alpinus* L., in Windermere, 1939–1982. *Journal of Fish Biology* 24, 229-234. **Kipling, C. & Frost, W.E. (1969).** Variations in the fecundity of Pike *Esox lucius* L. in Windermere. *Journal of Fish Biology* 1, 221-237. **Knezevic, B. (1981).** *Pseudorasbora parva*, new genus and species in the Lake Skadar. *Glas. Repl. Zavoda Zast. Prirode-Prirod. Muzeja.* 14, 79-84. **Kolomin, Y.M. (1977).** The Nadym River Ruffe, *Acerina cernua*. *Journal of Ichthyology* 17, 345-349. **Korwin-Kossakowski, M. (1988).** Larval development of carp, *Cyprinus carpio* L., in acidic water. *Journal of Fish Biology* 32, 17-26. **Kottelat, M. (1997).** European freshwater fishes – an heuristic checklist of the freshwater fishes of Europe (exclusive of former USSR), with an introduction for non-systematists and comments on nomenclature and conservation. *Biologia* 52, 1-271. **Kranenbarg, J., Winter, H.V. & Back, J.J.G.M. (2002).** Recent increase of North Sea houting *Coregonus oxyrhynchus* (L.) and prospects for recolonisation in the Netherlands. *Journal of Fish Biology* 61A, 251-253. **Krasznai, Z. & Marian, T. (1986).** Stock-induced triploidy and its effects on growth and gonad development of the European catfish, *Silurus glanis* L. *Journal of Fish Biology* 29, 519-528. **Krause, J. (1993).** The influence of hunger on shoal size by three-spined sticklebacks, *Gasterosteus aculeatus*. *Journal of Fish Biology* 43, 775-780. **Krause, J., Stoaks, G. & Mehner, T. (1998).** Habitat choice in shoals of roach as a function of water temperature and feeding rate. *Journal of Fish Biology* 53, 377-386.

L'Abee-Lund, J.H.L. (1988). Otolith shape discriminates between juvenile Atlantic salmon, *Salmo salar* L., and brown trout, *Salmo trutta* L. *Journal of Fish Biology* 33, 899-904. **Ladich, F. (1988).** Sound production by the gudgeon, *Gobio gobio* L., a common European freshwater fish (Cyprinidae, Teleostei). *Journal of Fish Biology* 32, 707-716. **Ladich, F. (1989).** Sound production by the river bullhead, *Cottus gobio* L., (Cottidae, Teleostei). *Journal of Fish Biology* 35, 531-538. **Lammens, E.H.R.R. & Hoogenboezem, W. (1991).** Diets and feeding behaviour. Pages 353-376 in: Winfield, I.J. & Nelson, J.S. (Eds): *Cyprinid Fishes: Systematics, Biology and Exploitation*. Fish and Fisheries Series

3, Chapman & Hall. **Lang, C. (1987).** Mortality of perch, *Perca fluviatilis* L., estimated from the size and abundance of egg strands. *Journal of Fish Biology* 31, 715-720. **Lawler, G.H. (1963).** The biology and taxonomy of the Burbot, *Lota lota*, in Hemming Lake, Manitoba. *Journal of the Fisheries Research Board of Canada* 20, 417-433. **Le Cren, E.D. (1985).** *The biology of the sea trout: summary of a symposium held at Plas Menai, 24-26 October, 1984.* Atlantic Salmon Trust, Pitlochry. **Le Cren, E.D., Kipling, C. & McCormack, J. (1967).** A study of the numbers, biomass and year class strengths of Perch (*Perca fluviatilis* L.) in Windermere from 1941–1966. *Journal of Animal Ecology* 46, 281-307. **Leeming, J.B. (1964).** The Chub, Bream and other fishes of the Welland. *Proceedings of the British Coarse Fish Conference, Liverpool* 1, 48-52. **Leeming, J.B. (1970).** A note on the occurrence of unusual mortalities amongst Common Carp (*Cyprinus carpio*). *Fisheries Management* 1, 9. **Lees, S. & Whitfield, P.J. (1992).** Virus-associated spawning papillomastosis in smelt, *Osmerus eperlanus* L., in the River Thames. *Journal of Fish Biology* 40, 503-510. **Lein, L. (1981).** Biology of the Minnow *Phoxinus phoxinus* and its interactions with Brown Trout *Salmo trutta* in Ovre Heimdalsvatn, Norway. *Holarctic Ecology* 4, 191-200. **Lelek, A. (1973).** Occurrence of the Sea Lamprey in midwater off Europe. *Copeia* 1, 136-137. **Lepage, M. & Rochard, E. (1995).** Threatened fishes of the world: *Acipenser sturio* Linnaeus, 1758 (Acipenseridae). *Environmental Biology of Fishes* 43, 28. **Lever, C. (1977).** *The Naturalised Animals of the British Isles.* Hutchinson, London. **Lever, C. (1997).** *Naturalized Fishes of the World.* Academic Press, London. **Levesley, P.B. & Magurran, A.E. (1988).** Population differences in the reaction of minnows to alarm substance. *Journal of Fish Biology* 32, 699-706. **Lewis, C. (1965).** *The Largemouth Bass fishery, Lake Opinicon, Ontario.* Unpublished MSc Thesis, University of Kingston. **Lewis, D.B., Walkey, M. & Dartnell, H.J.G. (1972).** Some effects of low oxygen tensions on the distribution of the Three-spined Stickleback *Gasterosteus aculeatus* L. and the Nine-spined Stickleback *Pungitius pungitius* (L.). *Journal of Fish Biology* 4, 103-108. **Linfield, R.S.J. (1979).** Changes in the rate of growth in a stunted Roach *Rutilus rutilus* population. *Journal of Fish Biology* 15, 275-298. Linfield, R.S.J. & Rickards, R.B. (1979). Zander in perspective. *Fisheries Management* 10, 1-16. **Lodi, E. (1967).** Sex reversal of *Cobitis taenia* L. (Osteichthyes, Fam. Cobitidae). *Experiantia* 23, 446. **Lodi, E. (1980).** Hermaphroditic and gonochoric populations of *Cobitis taenia bilineata* Canestrini (Cobitidae Osteichthyes). *Monitore Zoologia Italia* 14, 235-243. **Lohnisky, K. (1966).** The spawning behaviour of the Brook Lamprey, *Lampetra planeri* (Bloch, 1784). *Vestnik Ceskoslovenske Spolecnosti Zoolicke* 30, 289-307. **Lorenzen, K., des Clers, S.A. & Anders, K. (1991).** Population dynamics of lymphocystis disease in estuarine flounder, *Platichthys flesus* (L.). *Journal of Fish Biology* 39, 577-588. **Lyle, A.A. & Maitland, P.S. (1992).** Conservation of freshwater fish in the British Isles: the status of fish in National Nature Reserves. *Aquatic Conservation* 2, 19-34. **Lyle, A.A., Maitland, P.S. & Sweetman, K.E. (1996).** The spawning migration of the smelt *Osmerus eperlanus* in the River Cree, S.W. Scotland. *Biological Conservation* 80, 303-311.

MacCrimmon, H.R. (1971). World distribution of Rainbow Trout (*Salmo gairdneri*). *Journal of the Fisheries Research Board of Canada* 28, 663-704.

MacCrimmon, H.R. & Campbell, J.S. (1969). World distribution of Brook Trout, *Salvelinus fontinalis*. *Journal of the Fisheries Research Board of Canada* 26, 1699-1725. **MacCrimmon, H.R. & Devitt, O.E. (1954).** Winter studies on the Burbot, *Lota lota lacustris*, of Lake Simcoe, Ontario. *Canadian Fish Culturalist* 16, 34-41. **MacCrimmon, H.R. & Gots, B.L. (1979).** World distribution of Atlantic Salmon, *Salmo salar*. *Journal of the Fisheries Research Board of Canada* 36, 422-457. **MacCrimmon, H.R. & Marshall, T.L. (1968).** World distribution of Brown Trout, *Salmo trutta*. *Journal of the Fisheries Research Board of Canada* 25, 2527-2548. **MacDonald, T.H. (1959a).** Estimates of length of larval life in three species of lamprey found in Britain. *Journal of Animal Ecology* 28: 293-298. **MacDonald, T.H. (1959b).** Identification of ammocoetes of British lampreys. *Glasgow Naturalist* 18: 91-95. **Mackay, D.W. (1970).** Populations of Trout and Grayling in two Scottish rivers. *Journal of Fish Biology* 2, 39-45. **Magnhagen, C. (1998).** Alternative reproductive tactics and courtship in the common goby. *Journal of Fish Biology* 53, 130-137. **Magurran, A.E. (1986).** The development of shoaling behaviour in the European Minnow, *Phoxinus phoxinus*. *Journal of Fish Biology* 29A, 159-169. **Maitland, P.S. (1965).** The feeding relationships of Salmon, Trout, Minnows, Stone Loach and Three-spined Sticklebacks in the River Endrick, Scotland. *Journal of Animal Ecology* 34, 109-133. **Maitland, P.S. (1966a).** Present status of known populations of the Vendace, *Coregonus vandesius* Richardson, in Great Britain. *Nature, London*. 210, 216-217. **Maitland, P.S. (1966b).** The fish fauna of the Castle and Mill Lochs, Lochmaben, with special reference to the Lochmaben Vendace, *Coregonus vandesius* Richardson. *Transactions of the Dumfriesshire and Galloway Natural History and Antiquarian Society* 43, 31-48. **Maitland, P.S. (1967a).** The artificial fertilisation and rearing of the eggs of *Coregonus clupeoides* Lacepede. *Proceedings of the Royal Society of Edinburgh* 70, 82-106. **Maitland, P.S. (1967b).** Echo sounding observations on the Lochmaben Vendace, *Coregonus vandesius* Richards. *Transactions of the Dumfriesshire and Galloway Natural History and Antiquarian Society* 44, 29-46. **Maitland, P.S. (1969a).** A preliminary account of the mapping of the distribution of freshwater fish in the British Isles. *Journal of Fish Biology* 1, 45-58. **Maitland, P.S. (1969b).** The reproduction and fecundity of the Powan, *Coregonus clupeoides* Lacepede, in Loch Lomond, Scotland. *Proceedings of the Royal Society of Edinburgh* 70, B, 233-264. **Maitland, P.S. (1970).** The origin and present distribution of *Coregonus* in the British Isles. *International Symposium on the Biology of Coregonid Fishes, Winnipeg* 1, 99-114. **Maitland, P.S. (1971).** A population of coloured Goldfish, *Carassius auratus*, in the Forth and Clyde Canal. *Glasgow Naturalist* 18, 565-568. **Maitland, P.S. (1972a).** A key to the freshwater fishes of the British Isles. *Scientific Publications of the Freshwater Biological Association* 27, 1-137. **Maitland, P.S. (1972b).** Loch Lomond: man's effects on the salmonid community. *Journal of the Fisheries Research Board of Canada* 29, 849-860. **Maitland, P.S. (1979).** The status and conservation of rare freshwater fishes in the British Isles. *Proceedings of the British Coarse Fish Conference, Liverpool* 1, 237-248. **Maitland, P.S. (1980a).** Review of the ecology of lampreys in northern Europe. *Canadian Journal of Fisheries and Aquatic Sciences* 37, 1944-1952. **Maitland, P.S. (1980b).** Scarring of Whitefish (*Coregonus lavaretus*) by European River Lamprey (*Lampetra fluviatilis*) in Loch Lomond, Scotland. *Canadian Journal of Fisheries and Aquatic Sciences*

37, 1981-1988. **Maitland, P.S. (1982).** Elusive lake fish. *Living Countryside* 7, 1672-1673. **Maitland, P.S. (1983).** Catfishes: fish with 'whiskers'. *Living Countryside* 10, 2212-2213. **Maitland, P.S. (1990).** *The Biology of Fresh Waters*. Second Edition. Glasgow: Blackie. **Maitland, P.S. (1991).** Climate change and fish in northern Europe: some possible scenarios. *Proceedings of the Institute of Fishery Management, Annual Study Course* 22, 97-110. **Maitland, P.S. (2000).** *Guide to Freshwater Fish of Britain and Europe*. Hamlyn, London. **Maitland, P.S. & Campbell, R.N. (1992).** *Freshwater Fishes of the British Isles*. HarperCollins, London. **Maitland, P.S. & East, K. (1989).** An increase in numbers of Ruffe, *Gymnocephalus cernua* (L.), in a Scottish loch from 1982 to 1987. *Aquaculture and Fisheries Management* 20, 227-228. **Maitland, P.S., East, K. & Morris, K.H. (1983).** Ruffe *Gymnocephalus cernua* (L.), new to Scotland, in Loch Lomond. *Scottish Naturalist* 1983, 7-9. **Maitland, P.S., Greer, R.B., Campbell, R.N. & Friend, G.F. (1984).** The status of the Arctic Charr, *Salvelinus alpinus* (L.), in Scotland. *International Symposium on Arctic Charr, Winnipeg* 1, 193-215. **Maitland, P.S. & Lyle, A.A. (1991).** Conservation of freshwater fish in the British Isles: the current status and biology of threatened species. *Aquatic Conservation* 1, 25-54. **Maitland, P.S. & Lyle, A.A. (1996).** The smelt in Scotland. *Freshwater Forum* 6, 57-68. **Maitland, P.S., Morris, K.H. & East, K. (1994).** The ecology of lampreys (Petromyzonidae) in the Loch Lomond area. *Hydrobiologia* 290, 105-120. **Maitland, P.S., Morris, K.H., East, K., Schoonoord, M.P., Van der Wal, B. & Potter, I.C. (1984).** The estuarine biology of the River Lamprey, *Lampetra fluviatilis*, in the Firth of Forth, Scotland, with particular reference to size composition and feeding. *Journal of Zoology, London* 203, 211-225. **Maitland, P.S. & Price, C.E. (1969).** *Urocleidus principalis*, a North American monogenetic trematode new to the British Isles, probably introduced with the Largemouth Bass *Micropterus salmoides*. *Journal of Fish Biology* 1, 17-18. **Malloch, P.D. (1910).** *Life History of the Salmon, Sea Trout and other Freshwater Fish*. Black, London. **Malmqvist, B. (1978).** Population structure and biometry of *Lampetra planeri* (Bloch) from three different watersheds in south Sweden. *Archiv für Hydrobiologie* 84, 65-86. **Malmqvist, B. (1980a).** Habitat selection of larval Brook Lampreys (*Lampetra planeri*, Bloch) in a south Swedish stream. *Oecologia* 45, 35-38. **Malmqvist, B. (1980b).** The spawning migration of the Brook Lamprey, *Lampetra planeri* Bloch, in a south Swedish stream. *Journal of Fish Biology* 16, 105-114. **Manion, P.J. (1967).** Diatoms as food of larval Sea Lampreys in a small tributray of northern Lake Michigan. *Transactions of the American Fisheries Society* 96, 224-226. **Manion, P.J. & McLain, A.L. (1971).** Biology of larval sea lampreys (*Petromyzon marinus*) of the 1960 year class, isolated in the Big Garlic River, Michigan, 1960-65. *Great Lakes Fishery Commission Technical Report* 16, 1-35. **Mann, R.H.K. (1973).** Observations on the age, growth, reproduction and food of the Roach *Rutilus rutilus* (L.) in two rivers in southern England. *Journal of Fish Biology* 5, 707-736. **Mann, R.H.K. (1974).** Observations on the age, growth, reproduction and food of the Dace, *Leuciscus leuciscus* (L.) in two rivers in southern England. *Journal of Fish Biology* 6, 237-253. **Mann, R.H.K. (1976a).** Observations on the age, growth, reproduction and food of the Pike, *Esox lucius* (L.) in two rivers in southern England. *Journal of Fish Biology* 8, 179-197. **Mann, R.H.K. (1976b).** Observations on the age, growth, reproduction and food of the Chub *Squalius* *cephalus* (L.) in the River Stour, Dorset. *Journal of Fish Biology* 8, 265-288. **Mann, R.H.K. (1980).** The growth and reproductive strategy of the Gudgeon *Gobio gobio* (L.) in two hard-water rivers in southern England. *Journal of Fish Biology* 17, 163-176. **Mann, R.H.K. (1982).** The annual food consumption and prey preferences of Pike (*Esox lucius*) in the River Frome, Dorset. *Journal of Animal Ecology* 51, 81-95. **Mann, R.H.K. & Mills, C.A. (1983).** Biological and climatic influences on the dace *Leuciscus leuciscus* in a southern chalk stream. *Report of the Freshwater Biological Association* 54, 123-126. **Mann, R.H.K. & Steinmetz, B. (1985).** On the accuracy of age-determination using scales from rudd, *Scardinius erythrophthalmus* (L.), of known age. *Journal of Fish Biology* 26, 621-628. **Mansfield, K. (1958).** Pike-perch in England. *Salmon and Trout Magazine* 153, 94-98. **Marconato, A. & Bisazza, A. (1988).** Mate choice, egg cannibalism and reproductive success in the river bullhead, *Cottus gobio* L. *Journal of Fish Biology* 33, 905-916. **Marlborough, D. (1966).** The reported distribution of the crucian carp in Britain, 1954 to 1962. *Naturalist* 89, 1-3. **Marlborough, D. (1970).** The status of the Burbot, *Lota lota* (L.) (Gadidae) in Britain. *Journal of Fish Biology* 2, 217-222. **Martinez, A.M. (1984).** Identification of brook, brown, rainbow, and cutthroat trout larvae. *Transactions of the American Fisheries Society* 113, 252-259. **Masterman, A.T. (1913).** Report on investigations upon the smelt (*Osmerus eperlanus*) with special reference to age determination by study of scales, and its bearing upon sexual maturity. *Fisheries Investigation Series* 1, 113-126. **Mathews, C.P. & Williams, W.P. (1972).** Growth and annual check formation in scales of Dace, *Leuciscus leuciscus*. *Journal of Fish Biology* 4, 363-368. **Maxwell, H.C. (1904).** *British Freshwater Fishes*. Hutchinson, London. **Mayer, I., Shackley, S.E. & Witthames, P.R. (1988).** Aspects of the reproductive biology of the bass, *Dicentrarchus labrax* L. I. An histological and histochemical study of oocyte development. *Journal of Fish Biology* 33, 609-622. **Mayer, I., Shackley, S.E. & Witthames, P.R. (1990).** Aspects of the reproductive biology of the bass, *Dicentrarchus labrax* L. II. Fecundity and pattern of oocyte development. *Journal of Fish Biology* 36, 141-148. **McKenzie, J.A. & Keenleyside, M.H.A. (1970).** Reproductive behaviour of Ninespine Sticklebacks (*Pungitius pungitius* (L.)) in South Bay, Manitoulin Island, Ontario. *Canadian Journal of Zoology* 48, 55-61. **McLeave, J.D. (1980).** Swimming performance of the European eel, *Anguilla anguilla* (L.) elvers. *Journal of Fish Biology* 16, 445-452. **McPhail, J.D. (1966).** The *Coregonus autumnalis* complex in Alaska and northwestern Canada. *Journal of the Fisheries Research Board of Canada* 23, 141-148. **Meadows, B.S. (1968).** On the occurrence of the Guppy *Lebistes reticulatus* in the River Lee. *Essex Naturalist* 32, 186-189. **Menneson-Boisneau, C., Boisneau, P. & Bagliniere, J.L. (1986).** Premieres observations sur les caracteristiques biologiques des adultes de grande alose (*Alosa alosa* L.) dans le cours moyen de la Loire. *Acta Oecologia* 7, 337-353. **Miller, P.J. (1975).** Age structure and life span in the Common Goby *Pomatoschistus microps*. *Journal of Zoology, London* 177, 425-448. **Miller, P.J. & Loates, M.J. (1997).** *Fish of Britain and Europe*. HarperCollins, London. **Mills, C.A. (1980).** Spawning and rearing eggs of the dace, *Leuciscus leuciscus* (L.). *Fisheries Management* 11, 67-72. **Mills, C.A. (1982).** Factors affecting the survival of dace, *Leuciscus leuciscus* (L.), in the early post-hatching period. *Journal of Fish Biology* 20, 645-656. **Mills, C.A. (1988).** The effect of extreme northerly climatic conditions on the life history of the minnow, *Phoxinus phoxinus* (L.). *Journal of Fish Biology* 33, 545-562. **Mills, C.A., Welton, J.S & Rendle, E.L. (1983).** The age, growth and reproduction of the Stone Loach *Noemacheilus barbatulus* (L.) in a Dorset chalk stream. *Freshwater Biology* 13, 283-292. **Mills, D.H. (1964).** The ecology of the young stages of the Atlantic Salmon in the River Bran, Ross-shire. *Freshwater and Salmon Fisheries Research* 32, 1-58. **Mills, D.H. (1969).** The growth and population densities of Roach in some Scottish waters. *Proceedings of the British Coarse Fish Conference, Liverpool* 4, 50-57. **Mills, D.H. (1971).** *Salmon and Trout: A Resource, its Ecology, Conservation and Management.* Oliver and Boyd, Edinburgh. **Milner, N.J., Gee, A.S. & Hemsworth, R.J. (1978).** The production of Brown Trout, *Salmo trutta* in tributaries of the upper Wye, Wales. *Journal of Fish Biology* 13, 599-612. **Mooij, W.M. (1989).** Key to the identification of larval bream, white bream and roach. *Journal of Fish Biology* 34, 111-118. **Moore, J.W. & Moore, I.A. (1976).** The basis of food selection in Flounders, *Platichthys flesus* (L.), in the Severn Estuary. *Journal of Fish Biology* 9, 139-156. **Moore, J.W. & Potter, I.C. (1976).** Aspects of feeding and lipid deposition and utilisation in the lampreys, *Lampetra fluviatilis* (L.) and *Lampetra planeri* (Bloch). *Journal of Animal Ecology* 45, 699-712. **Moriarty, C. (1973).** Studies of the Eel, *Anguilla anguilla*, in Ireland: 2. In Lough Conn, Lough Gill and North Cavan Lakes. *Irish Fisheries Investigations* 13, 1-13. **Moriarty, C. (1978).** *Eels*. David and Charles, Newton Abbot. **Moriarty, C. (1983).** Age determination and growth of Eels, *Anguilla anguilla. Journal of Fish Biology* 23, 257-264. **Morman, R.H., Cuddy, D.W. & Rugen, P.C. (1980).** Factors influencing the distribution of Sea Lamprey (*Petromyzon marinus*) in the Great Lakes. *Canadian Journal of Fisheries and Aquatic Sciences* 37, 1811-1826. **Morris, D. (1952).** Homosexuality in the Ten-spined Stickleback (*Pygosteus pygosteus* L.). *Behaviour* 4, 233-261. **Morris, K.H. (1978).** The food of the Bullhead (*Cottus gobio* L.) in the Gogar Burn, Lothian, Scotland. *Forth Naturalist and Historian* 7, 31-44. **Morris, K.H., (1989).** A multivariate morphometric and meristic description of a population of freshwater-feeding River Lampreys, *Lampetra fluviatilis* (L.), from Loch Lomond, Scotland. *Zoological Journal of the Linnaean Society of London* 96: 357-371. **Mraz, D., Kmiotek, S. & Frankenberger, L. (1961).** The Largemouth Bass. Its life history, ecology and management. *Wisconsin Conservation Department Publication* 232, 1-13. **Mulicki, Z. (1947).** The food and feeding habit of the Flounder (*Pleuronectes flesus* L.) in the Gulf of Gdansk. *Archiwum Hydrobiologji i. Rybactwa.* 13, 221-259. **Munro, M.A., Whitfield, P.J. & Diffley, R. (1989).** *Pomphorhynchus laevis* (Muller) in the flounder, *Platichthys flesus* L., in the tidal River Thames: population structure, microhabitat utilization and reproductive status in the field and under conditions of controlled salinity. *Journal of Fish Biology* 35, 719-736. **Munro, W.R. (1957).** The Pike of Loch Choin. *Freshwater and Salmon Fisheries Research* 16, 1-16. **Myers, R.A. & Hutchings, J.A. (1987).** Mating of anadromous Atlantic salmon, *Salmo salar* L., with mature male parr. *Journal of Fish Biology* 31, 143-146.

Naesje, T.F., Jonsson, B., Klyve, L. & Sandlund, O.T. (1987). Food and growth of age-0 Smelts, *Osmerus eperlanus*, in a Norwegian fjord lake. *Journal of Fish Biology* 30, 119-126. **Naish, K.A. Carvalho, G.R. & Pitcher, T.J. (1993).** The genetic structure and microdistribution of shoals of *Phoxinus*

phoxinus, the European minnow. *Journal of Fish Biology* 43A, 75-90. **Naismith, I.A. & Knights, B. (1990).** Modelling of unexploited and exploited populations of eels, *Anguilla anguilla* (L.), in the Thames Estuary. *Journal of Fish Biology* 37, 975-986. **Naismith, I.A. & Knights, B. (1993).** The distribution, density and growth of the European eel, *Anguilla anguilla*, in the freshwater catchment of the River Thames. *Journal of Fish Biology* 42, 217-226. **Nall, G.H. (1930).** *The Life of the Sea Trout.* Seeley Service, London. **Narver, D.W. (1969).** Age and size of steelhead trout in Babine River, British Columbia. *Journal of the Fisheries Research Board of Canada* 26, 2754-2760. **Nellbring, S. (1989).** The ecology of smelts (Genus *Osmerus*): a literature review. *Nordic Journal of Freshwater Research* 65, 116-145. **Nelson, J.S. (1994).** *Fish of the World.* Wiley, New York. **Newth, H.G. (1930).** The feeding of ammocoetes. *Nature, London* 126, 94-95. **Nilsson, G.E. (1990).** Distribution of aldehyde dehydrogenase and alcohol dehydrogenase in summer-acclimatized crucian carp, *Carassius carassius* L. *Journal of Fish Biology* 36, 175-180. **Noltie, D.B. & Keenleyside, M.H.A. (1986).** Correlates of reproductive success in stream-dwelling male rock bass, *Ambloplites rupestris* (Centrarchidae). *Environmental Biology of Fishes* 17, 61-70.

O'Grady, K.T. & Spillett, P.B. (1985). A comparison of the pond culture of carp, *Cyprinus carpio* L., in Britain and mainland Europe. *Journal of Fish Biology* 26, 701-714. **O'Hara, J.J. (1968).** Influence of weight and temperature on metabolic rate of sunfish. *Ecology* 1, 159-161. **Ojutkangas, E. Aronen, K. & Laukkanen, E. (1995).** Distribution and abundance of river lamprey (*Lampetra fluviatilis*) ammocoetes in the regulated River Perhonjoki. *Regulated Rivers: Research and Management* 10, 239-245. **Oliva, O. & Vostradovsky, J. (1960).** Contribution to the knowledge of growth of the Pope *Acerina cernua. Casopis Narodniho Musea.* 129, 56-63. **O'Maoileidigh, N.O., Cawdery, S., Bracken, J.J. & Ferguson, A. (1988).** Morphometric, meristic character and electrophoretic analyses of two Irish Populations of twaite shad, *Alosa fallax* (Lacepede). *Journal of Fish Biology* 32, 355-366. **Owen, G. (1955).** The use of propylene phenoxetol as a relaxing agent. *Nature, London* 175, 434.

Palliainen, E. & Korhonen, K. (1990). Seasonal changes in condition indices in adult mature and non-maturing burbot, *Lota lota* (L.), in the north-eastern Bothnian Bay, northern Finland. *Journal of Fish Biology* 36, 251-260. **Palliainen, E. & Korhonen, K. (1993).** Does the burbot, *Lota lota*, have rest years between normal spawning seasons? *Journal of Fish Biology* 43, 355-362. **Palmer, C.J. & Culley, M.B. (1984).** The egg and early life stages of the sandsmelt, *Atherina presbyter* Cuvier. *Journal of Fish Biology* 24, 537-544. **Papadopol, M. (1960).** Beitrage zur Kenntis der morphologischen Veranderung und der Biologie des Bitterlings (*Rhodeus sericeus amarus* - Bloch). *Biologische Studien und Firschungen. Serie Zoologie* 12, 81-190. **Parkinson, D., Philippart, J.C. & Baras, E. (1999).** Preliminary investigation of spawning migrations of grayling in a small stream as determined by radio-tracking. *Journal of Fish Biology* 55, 172-182. **Partington, J.D. & Mills, C.A. (1988).** An electrophoretic and biometric study of Arctic charr, *Salvelinus alpinus* (L.), from ten British lakes. *Journal of Fish Biology* 33, 791-814. **Pawson, M.G. & Eaton, D.R. (1999).** The influence of a power station on the survival of juvenile sea bass in an estuarine nursery area. *Journal of Fish Biology* 54, 1143-1160. **Pawson, M.G., Kelley, D.F. & Pickett, G.D. (1987).** The distribution and migrations of Bass, *Dicentrarchus labrax* L., in waters around England and Wales as shown by tagging. *Journal of the Marine Biological Association of the UK* 67, 263-274. **Pawson, M.G. & Pickett, G.D. (1987).** *The Bass (*Dicentrarchus labrax*) and Management of its Fishery in England and Wales.* MAFF, Lowestoft. **Payne, R.H., Child, A.R. & Forrest, A. (1971).** The existence of natural hybrids between European Trout and the Atlantic Salmon. *Journal of Fish Biology* 4, 233-236. **Penaz, M. (1968).** Development of the chub, *Leuciscus cephalus* (Linnaeus 1758) in the post hatching period. *Zoologicke Listy* 17, 269-278. **Perez, J., Martinez, J.L., Moran, P., Beall, E. & Garcia-Vasquez, E. (1999).** Identification of Atlantic salmon and brown trout hybrids with a nuclear marker useful for evolutionary studies. *Journal of Fish Biology* 54, 460-464. **Petridis, D. (1990).** The influence of grass carp on habitat structure and its subsequent effect on the diet of tench. *Journal of Fish Biology* 36, 533-544. **Petrimoulx, H.J. (1984).** Observations on the spawning behaviour of the Roanoke bass. *Progressive Fish-Culturalist* 46,120-125. **Philipp, D.P., Childers, W.F. & Whitt, G.S. (1985).** Correlations of allele frequencies with physical and environmental variables for populations of largemouth bass, *Micropterus salmoides* (Lacepede). *Journal of Fish Biology* 27, 347-366. **Phillips, M.J. (1989).** The feeding sounds of rainbow trout, *Salmo gairdneri* Richardson. *Journal of Fish Biology* 35, 589-592. **Pilcher, M.W. & Moore, J.F. (1993).** Distribution and prevalence of *Anguillicola crassus* in eels from the tidal Thames catchment. *Journal of Fish Biology* 43, 339-344. **Pinder, A.C. (2001).** Keys to larval and juvenile stages of coarse fishes from fresh waters in the British Isles. *Scientific Publications of the Freshwater Biological Association* 50, 1-136. **Pitcher, T.J., Green, D.A. & Magurran, A.E. (1986).** Dicing with death: predator inspection behaviour in Minnow shoals. *Journal of Fish Biology* 28, 439-448. **Pitts, C.S., Jordan, D.R., Cowx, I.G. & Jones, N.V. (1997).** Controlled breeding studies to verify the identity of roach and common bream hybrids from a natural population. *Journal of Fish Biology* 51, 686-696. **Poncin, P. (1992).** Influence of the daily distribution of light on reproduction in the barbel, *Barbus barbus* (L.). *Journal of Fish Biology* 41, 993-998. **Potter, I.C. & Osborne, T.S. (1975).** The systematics of British larval lampreys. *Journal of Zoology, London* 176, 311-329. **Pope, J.A., Mills, D.H. & Shearer, W.M. (1961).** The fecundity of Atlantic salmon (*Salmo salar* Linn.). *Freshwater Salmon Fishery Research, Scotland* 26, 1-12. **Potter, I.C. (1980).** Ecology of larval and metamorphosing lampreys. *Canadian Journal of Fisheries and Aquatic Sciences* 37, 1641-1657. **Pottinger, T.G. (1998).** Changes in blood cortisol, glucose and lactate in carp retained in anglers' keepnets. *Journal of Fish Biology* 53, 728-742. **Probst, W.E., Rabeui, C.F., Covington, W.G. & Marteney, R.E. (1984).** Resource use by stream-dwelling rock bass and smallmouth bass. *Transactions of the American Fisheries Society* 113, 283-294. **Prouzet, P., Martinet, J.P. & Badia, J. (1994).** Caracterisation biologique et variation des captures de la grande alose (*Alosa alosa*) par unite d'effort sur le fleuve Adour (Pyrenees Atlantiques, France). *Aquatic Living Resources* 7, 1-10. **Puke, C. (1952).** Pike-perch studies in Lake Vanern. *Report of the Institute of Freshwater Research, Drottningholm* 33, 168-178.

Pyefinch, K.A. (1955). A review of the literature on the biology of the Atlantic salmon (*Salmo salar* Linn.). *Freshwater Salmon Fisheries Research, Scotland* 9, 1-24.

Radforth, I. (1940). The food of the Grayling (*Thymallus thymallus*), Flounder (*Platichthys flesus*), Roach (*Rutilus rutilus*) and Gudgeon (*Gobio fluviatilis*), with special reference to the Tweed watershed. *Journal of Animal Ecology* 9, 302-318. **Raicu, P., Taisescu, E. & Banarescu, P. (1981).** *Carassius carassius* and *C. auratus*, a pair of diploid and tetraploid representative species (Pisces, Cyprinidae). *Cytologia* 46, 233-240. **Rask, M., Vuorinen, P.J. & Vuorinen, M. (1990).** Delayed spawning of perch, *Perca fluviatilis* L., in acidified lakes. *Journal of Fish Biology* 36, 317-326. **Rasottos, M.B., Cardellini, P. & Marconato, E. (1987).** The problem of sexual inversion in the Minnow, *Phoxinus phoxinus*. *Journal of Fish Biology* 30, 51-58. **Reay, P.J. (1992).** *Mugil cephalus* L. – a first British record and a further 5°N. *Journal of Fish Biology* 40, 311-313. **Reay, P.J. & Cornell, V. (1988).** Identification of grey mullet (Teleostei: Mugilidae) juveniles from British waters. *Journal of Fish Biology* 32, 95-100. **Regan, C.T. (1911).** *The Freshwater Fishes of the British Isles*. Methuen, London. **Reid, H. (1930).** A study of *Eupotomis gibbosus* (L.) as occurring in the Chamcook Lakes, N.B. *Contributions to Canadian Biology of Fish* 5, 457-466. **Reiersen, L.O. & Fugelie, K. (1984).** Annual variation in lymphocystis infection frequency in flounder, *Platichthys flesus* (L.). *Journal of Fish Biology* 24, 187-192. **Reimers, N. (1979).** A history of a stunted Brook Trout population in an alpine lake: a lifespan of 24 years. *California Fish and Game* 64, 196-215. **Rickards, B. & Fickling, N. (1979).** *Zander*. Black, London. **Roberts, R.J., Leckie, J. & Slack, H.D. (1970).** Bald spot disease in powan. *Journal of Fish Biology* 2, 103-105. **Robinson, G.D., Dunson, W.A., Wright, J.E. & Mamolito, G.E. (1976).** Differences in low pH tolerance among strains of brook trout (*Salvelinus fontinalis*). *Journal of Fish Biology* 8, 5-17. **Robotham, P.W.J. (1977).** Feeding habits and diet in two populations of Spined Loach, *Cobitis taenia* (L.). *Freshwater Biology* 7, 469-477. **Robotham, P.W.J. (1982a).** An analysis of a specialised feeding mechanism of the Spined Loach, *Cobitis taenia* (L.), and a description of the related structures. *Journal of Fish Biology* 20, 173-181. **Robotham, P.W.J. (1982b).** Infection of the Spined Loach, *Cobitis taenia*, by the digenean, *Allocreadium transversale* (Rud.). *Journal of Fish Biology* 21, 699-703. **Rochard, E., Castelnaud, G. & Lepage, M. (1990).** Sturgeons (Pisces Acipenseridae): threats and prospects. *Journal of Fish Biology* 37A, 123-132. **Rochard, E., Lepage, M. & Meauze, L. (1997).** Identification et caracterisation de l'aire de repartition marine de l'esturgeon europeen *Acipenser sturio* a partir de declarations de captures. *Aquatic Living Resources* 10, 101-109. **Rogers, S.I. (1988).** Reproductive effort and efficiency in the female common goby, *Pomatoschistus microps* (Kroyer) (Teleostei: Gobioidei). *Journal of Fish Biology* 33, 109-120. **Romer, G.S. & McLachlan, A. (1986).** Mullet grazing on surf diatom accumulations. *Journal of Fish Biology* 28, 93-104. **Rosa, H. (1958).** *A Synopsis of Biological Data on Tench* Tinca tinca *(Linnaeus 1758)*. FAO Fisheries Division, Rome. **Rosecchi, E, Thomas, F. & Crivelli, A.J. (2001).** Can life-history traits predict the fate of introduced species? A case study on two cyprinid fish in southern France. *Freshwater Biology* 46, 845-853. **Rumpus, A.E. (1975).** The helminth parasites of the Bullhead, *Cottus gobio* (L.), and the Stone Loach, *Noemacheilus* *barbatulus* (L.) from the River Avon, Hampshire. *Journal of Fish Biology* 7, 469-483.

Sadler, K. (1979). Effect of temperature on the growth and survival of the European Eel, *Anguilla anguilla* L. *Journal of Fish Biology* 15, 499-507. **Salam, A. & Davies, P.M.C. (1994).** Effect of body weight and temperature on the maximum daily food consumption of *Esox lucius*. *Journal of Fish Biology* 44, 165-167. **Schindler, O. (1957).** *Freshwater Fishes*. Thames and Hudson, London. **Schmidt, J. (1922).** The breeding places of the eel. *Transactions of the Royal Society of London* 211B, 179-208. **Schwartz, F.J. (1972).** World literature to fish hybrids with an analysis by family, species and hybrids. *Publications of the Gulf Coast Research Laboratory Museum* 3, 1-328. **Scott, A. (1985).** Distribution, growth, and feeding of postemergent Grayling *Thymallus thymallus* in an English river. *Transactions of the American Fisheries Society* 114, 525-531. **Scott, D.B.C. (1975).** A hermaphrodite specimen of *Coregonus lavaretus* (L.) (Salmoniformes, Salmonidae) from Loch Lomond, Scotland. *Journal of Fish Biology* 7, 709. **Scott, D.C. (1949).** A study of a stream population of Rock Bass. *Investigations on Indiana Lakes and Streams* 3, 169-234. **Scott, W.B. & Crossman, E.J. (1973).** *Freshwater Fishes of Canada*. Fisheries Research Board of Canada, Ottawa. **Seaman, E.A. (1979).** Observations on *Carassius auratus* (Linnaeus) harvesting *Potamogeton foliosus* Raf. in a small pond in northern Virginia. *Fisheries* 4, 24-25. **Shafi, M. (1969).** *Comparative studies of populations of Perch (*Perca fluviatilis *L.) and Pike (*Esox lucius *L.) in two Scottish lochs.* Unpublished PhD Thesis, University of Glasgow. **Shafi, M. & Maitland, P.S. (1971a).** Comparative aspects of the biology of Pike *Esox lucius* in two Scottish lochs. *Proceedings of the Royal Society of Edinburgh* B, 71, 41-60. **Shafi, M. & Maitland, P.S. (1971b).** The age and growth of Perch *Perca fluviatilis* in two Scottish lochs. *Journal of Fish Biology* 3, 39-57. **Shearer, W.M. (1961).** Pacific Salmon in the North Sea. *New Scientist* 232, 184-186. **Sibbing, F.A. (1991).** Food capture and oral processing. Pages 377-412 in: Winfield, I.J. & Nelson, J.S. (Eds) *Cyprinid Fishes: Systematics, Biology and Exploitation*. Fish and Fisheries Series 3, Chapman & Hall. **Sibbing, F.A., Osse, J.W.M. & Terlouw, A. (1986).** Food handling in the carp (*Cyprinus carpio*): its movement patterns, mechanisms and limitations. *Journal of Zoology, London* (A) 210, 161-203. **Simon, T.P. & Vondruska, J.T. (1991).** Larval identification of the ruffe, *Gymnocephalus cernuus* (Linnaeus) (Percidae: Percini), in the St Louis River estuary, Lake Superior drainage system, Minnesota. *Canadian Journal of Zoology* 69, 436-442. **Sinha, V.R.P. & Jones, J.W. (1975).** *The European Freshwater Eel*. Liverpool University Press, Liverpool. **Sjoberg, K. (1980).** Ecology of the European River Lamprey (*Lampetra fluviatilis*) in northern Sweden. *Journal of the Fisheries Research Board of Canada* 37, 1974-1980. **Skryabin, A.G. (1993).** The biology of stone loach *Barbatula barbatula* in the Rivers Golonstanaya and Olkha, East Siberia. *Journal of Fish Biology* 42, 361-374. **Slack, H.D., Gervers, F.W.K. & Hamilton, J.D. (1957).** The biology of the Powan. *Glasgow University Publications, Studies on Loch Lomond* 1, 113-127. **Smagula, C.M. & Adelman, I.R. (1983).** Growth in a natural population of largemouth bass, *Micropterus salmoides* Lacepede, as determined by physical measurement and [14C]-glycine uptake by scales. *Journal of Fish Biology* 22, 695-704.

Smith, G.R. & Stearley, R.F. (1989). The classification and scientific names of rainbow and cutthroat trouts. *Fisheries* 14, 4-10. **Smith, I.R. & Lyle, A.A. (1979).** *Distribution of Freshwaters in Britain.* Institute of Terrestrial Ecology, Cambridge. **Smith, T.I. (1985).** The fishery, biology and management of Atlantic Sturgeon, *Acipenser oxyrhynchus*, in North America. Pages 61-72 in: Binkowski, F.P. & Doroshov, S.I. (Eds) *North American Sturgeons: Biology and Aquaculture Potential.* Junk, Dordrecht. **Smyly, W.J.P. (1955).** On the biology of the Stone-Loach *Nemacheilus barbatula*. *Journal of Animal Ecology* 24, 167-186. **Smyly, W.J.P. (1957).** The life-history of the Bullhead or Miller's Thumb (*Cottus gobio*). *Proceedings of the Zoological Society of London* 128, 431-453. **Solanki, T.G. & Benjamin, M. (1982).** Changes in the mucus cells of gills, buccal cavity and epidermis of the nine-spined stickleback, *Pungitius pungitius* L., induced by transferring to sea water. *Journal of Fish Biology* 21, 563- 575. **Solomon, D.J. & Child, A.R. (1978).** Identification of juvenile natural hybrids between Atlantic Salmon (*Salmo salar* L.) and Trout (*Salmo trutta* L.). *Journal of Fish Biology* 12, 499-502. **Soto, C.G., Zhang, J.S. & Shi, Y.H (1994).** Intraspecific cleaning behaviour in *Cyprinus carpio* in aquaria. *Journal of Fish Biology* 44, 172-174. **Spataru, P. (1967a).** Biologia nutritiei la zvirluga – *Cobitis taenia taenia* (Linnaeus 1758) – din compexul de balti Crapina-Jijila. *Analele Universitatii Bucuresti, Biologie* 16, 163-169. **Spataru, P. (1967b).** Locul obletului – *Alburnus alburnus alburnus* (Linnaeus 1758) in econimia complexului de Balti Crapina-Jijila (zona inundabila a Dunarii). *Buletinul Institutului de Cercetari Piscicole* 26, 68-76. **Spataru, P. (1968).** The nutrition of *Blicca bjoerkna bjoerkna* in the swamp complex Crapina-Jijila (Danube flooded area). *Hidrobiologia* 9, 219-226. **Spataru, P. & Gruia, L. (1967).** Die biologische Stellung des Bitterlings – *Rhodeus sericeus amarus* – im Flachseekomplex Crapina-Jijila. *Archiv für Hydrobiologie* 4, 420-432. **Spoor, W.A. (1977).** Oxygen requirements of embryos and larvae of the Large Mouth Bass, *Micropterus salmoides* (Lacepede). *Journal of Fish Biology* 11, 77-86. **Steffens, W. (1960).** Ernahrung und Wachstum des jungen Zanders (*Lucioperca lucioperca*). *Zeitschrift für Fischerei* 9, 161-271. **Stein, R.A. & Kitchell, J.F. (1975).** Selective predation by Carp (*Cyprinus carpio* L.) on benthic molluscs in Skadar Lake, Yugoslavia. *Journal of Fish Biology* 7, 391-399. **Stott, B. (1967).** The movements and population densities of roach (*Rutilus rutilus* (L.)) and gudgeon (*Gobio gobio* (L.)) in the River Mole. *Journal of Animal Ecology* 36, 407-423. **Stott, B, Elsdon, J.W.V. & Johnston, J.A.A. (1963).** Homing behaviour in gudgeon (*Gobio gobio*, (L.)). *Animal Behaviour* 11, 93-96. **Stott, B. & Buckley, B.R. (1979).** Avoidance experiments with homing shoals of Minnows, *Phoxinus phoxinus* in a laboratory stream channel. *Journal of Fish Biology* 14, 135-146. **Stott, B. & Cross, D.G. (1973).** A note on the effect of lowered temperatures on the survival of eggs and fry of the Grass Carp *Ctenopharyngodon idella* (Valenciennes). *Journal of Fish Biology* 5, 649-658. **Street, N.E. & Hart, P.J.B. (1985).** Group size and patch-location by the Stone Loach, *Noemacheilus barbatulus*, a non-visually foraging predator. *Journal of Fish Biology* 27, 785-792. **Street, N.E., Magurran, A.E. & Pitcher, T.J. (1984).** The effects of increasing shoaling size on handling time in goldfish, *Carassius auratus* L. *Journal of Fish Biology* 25, 561-566. **Stuart, T.A. (1953).** Spawning migration, reproduction and young stages of loch Trout (*Salmo trutta* L.). *Freshwater Salmon Fisheries Research* 5, 1-39. **Stuart, T.A. (1957).** The migrations and homing behaviour of brown trout (*Salmo trutta* L.). *Freshwater Salmon Fisheries Research, Scotland* 18, 1-27. **Summers, R.W. (1979).** Life cycle and population ecology of the Flounder *Platichthys flesus* (L.) in the Ythan Estuary, Scotland. *Journal of Natural History* 13, 703-723. **Sutela, T. & Huusko, A. (1997).** Food consumption of vendace *Coregonus albula* larvae in Lake Lentua, Finland. *Journal of Fish Biology* 51, 939-951. **Svardson, G. (1950).** Note on spawning habits of *Leuciscus erythrophthalmus*, *Abramis brama* and *Esox lucius*. *Annual Report of the Institute of Freshwater Fisheries Research, Drottningholm* 29, 102-107. **Svardson, G. (1956).** The coregonid problem. VI The palaearctic species and their intergrades. *Annual Report of the Institute of Freshwater Research, Drottningholm* 38, 267-356. **Svardson, G. & Molin, G. (1973).** The impact of climate on Scandinavian populations of the Sander, *Stizostedion lucioperca* (L.). *Annual Report of the Institute of Freshwater Fisheries Research, Drottningholm* 53, 112-139. **Sweeting, R.A. (1976).** Studies on *Ligula intestinalis* (L.) effects on a roach population in a gravel pit. *Journal of Fish Biology* 9, 515-522. **Swinney, G.N. & Coles, T.F. (1982).** Description of two hybrids involving Silver Bream *Blicca bjoerkna* from British UK waters. *Journal of Fish Biology* 20, 121-130.

Taverny, C. (1990). An attempt to estimate *Alosa alosa* and *Alosa fallax* juvenile mortality caused by three types of human activity in the Gironde Estuary, 1985-1986. *EIFAC Symposium, Goeteborg* 1988, 215-229. **Taverny, C. & Elie, P. (1988).** Mortalitees engendrees par l'industrie et la peche. Le cas des juveniles d'*Alosa alosa* et d'*Alosa fallax* dans l'estuaire de la Gironde en 1986. *CEMAGREF Publications.* 2, 1-52. **Taylor, J. & Mahon, R. (1977).** Hybridization of *Cyprinus carpio* and *Carassius auratus*, the first two exotic species in the lower Laurentian Great Lakes. *Environmental Biology of Fishes* 1, 205-208. **Thompson, B.M. & Harrop, R.T. (1987).** The distribution and abundance of Bass (*Dicentrarchus labrax*) eggs and larvae in the English Channel and southern North Sea. *Journal of the Marine Biological Association of the UK* 67, 263-274. **Thompson, D, & Iliadou, K. (1990).** A search for introgressive hybridization in the rudd, *Scardinius erythrophthalmus* (L.), and the roach, *Rutilus rutilus* (L.). *Journal of Fish Biology* 37, 367-374. **Thorpe, J.E. (1974).** Estimation of the number of Brown Trout *Salmo trutta* (L.) in Loch Leven, Kinross, Scotland. *Journal of Fish Biology* 6, 135-152. **Thorpe, J.E. (1977a).** Bimodal distribution of length of juvenile Atlantic Salmon (*Salmo salar* L.) under artificial rearing conditions. *Journal of Fish Biology* 11, 175-184. **Thorpe, J.E. (1977b).** Daily ration of adult perch, *Perca fluviatilis* L., during summer in Loch Leven, Kinross, Scotland. *Journal of Fish Biology* 11, 55-68. **Thorpe, J.E. (1987).** Smolting versus residency: developmental conflict in salmonids. *Symposium of the American Fisheries Society* 1, 244-252. **Tin, H.T. (1982a).** Family Ictaluridae, bullhead catfishes. Pages 436-457 in: Auer, N.A. (Ed.) *Identification of Larval Fishes of the Great Lakes Basin, with Emphasis on the Lake Michigan Drainage*. Great Lakes Fishery Commission Special Publication, Ann Arbor. **Tin, H.T. (1982b).** Family Centrarchidae, sunfishes. Pages 524-580 in: Auer, N.A. (Ed.) *Identification of Larval Fishes of the Great Lakes Basin, with Emphasis on the Lake Michigan Drainage*. Great Lakes Fishery Commission Special Publication, Ann Arbor. **Toner, E.D. (1959).** Predation by Pike (*Esox lucius* L.) in three Irish loughs. *Report on the Sea and*

Inland Fisheries of Ireland 25, 1-7. **Townsend, C.R. & Perrow, M.R. (1989).** Eutrophication may produce population cycles in roach; *Rutilus rutilus* (L.), by two contrasting methods. *Journal of Fish Biology* 34, 161-164. **Trautman, M.B. (1981).** *The Fishes of Ohio*. Ohio State University Press, Columbus. **Treasurer, J.W. (1976).** Age, growth and length-weight relationship of Brown Trout *Salmo trutta* (L.) in the Loch of Strathbeg, Aberdeenshire. *Journal of Fish Biology* 8, 241-253. **Treasurer, J.W. (1981).** Some aspects of the reproductive biology of Perch, *Perca fluviatilis* L., fecundity, maturation and spawning behaviour. *Journal of Fish Biology* 18, 729-740. **Treasurer, J.W. (1989).** Mortality and production of 0+ perch, *Perca fluviatilis* L., in two Scottish lakes. *Journal of Fish Biology* 34, 913-928. **Treasurer, J.W. (1990a).** The annual reproductive cycle of pike, *Esox lucius* L., in two Scottish lakes. *Journal of Fish Biology* 36, 29-48. **Treasurer, J.W. (1990b).** The occurrence of roach, *Rutilus rutilus* (L.), in northern Scotland. *Journal of Fish Biology* 37, 989-990. **Trewavas, E. (1938).** The Killarney Shad or Goureen (*Alosa fallax killarnensis*). *Proceedings of the Linnaean Society, London* 150, 110-112. **Tucker, D.W. (1959).** A new solution to the Atlantic eel problem. *Nature, London* 183, 495-501. **Tuunainen, P., Ikonen, E. & Auvinen, H. (1980).** Lamprey and lamprey fisheries in Finland. *Canadian Journal of Fisheries and Aquatic Sciences* 37, 1953-1959. **Twomey, E. (1956).** Pollan of Lough Erne. *Irish Naturalists' Journal* 12, 14-17. **Tyler, C.R. & Everett, S. (1993).** Incidences of gross morphological disorders in barbel (*Barbus barbus*) in three rivers in southern England. *Journal of Fish Biology* 43, 739-748.

Vadas, R.L. (1990). The importance of omnivory and predator regulation of prey in freshwater fish assemblages of North America. *Environmental Biology of Fishes* 27, 285-302. **Valente, A.C.N. (1988).** A note on fin abnormalities in *Leuciscus cephalus* L., and *Carassius carassius* L. (Pisces: Cyprinidae). *Journal of Fish Biology* 32, 633-634. **Valtonen, T. (1980).** European River Lamprey (*Lampetra fluviatilis*) fishing and lamprey populations in some rivers running into Bothnian Bay, Finland. *Canadian Journal of Fisheries and Aquatic Sciences* 37, 1967-1973. **Van den Broek, W.L.F. (1979).** Copepod ectoparasites of *Merlangius merlangus* and *Platichthys flesus*. *Journal of Fish Biology* 14, 371-380. **Van Dyke, J.M. & Sutton, D.L. (1977).** Digestion of Duckweed (*Lemna* spp.) by the Grass Carp (*Ctenopharyngodon idella*). *Journal of Fish Biology* 11, 273-278. **Van Eenennaam, J.P. & Doroshove, S.I. (1998).** Effects of age and body size on gonadal development of Atlantic sturgeon. *Journal of Fish Biology* 53, 624-637. **Verspoor, E. (1988).** Widespread hybridization between native Atlantic salmon, *Salmo salar*, and introduced brown trout, *Salmo trutta*, in eastern Newfoundland. *Journal of Fish Biology* 32, 327-334. **Vilizzi, L. (1998).** Age, growth and cohort composition of 0+ carp in the River Murray, Australia. *Journal of Fish Biology* 52, 997-1013. **Vollestad, L.A. & L'Abee-Lund, J.H. (1990).** Geographic variation in life history strategy of female roach, *Rutilus rutilus* (L.). *Journal of Fish Biology* 37, 853-864. **Vuorinen, J., Naesje, T.F. & Sandlund, O.T. (1991).** Genetic changes in a vendace, *Coregonus albula* (L.), population, 92 years after introduction. *Journal of Fish Biology* 39A, 193-202.

Walker, A.F., Greer, R.B. & Gardner, A.S. (1988). Two ecologically distinct forms of Arctic Charr (*Salvelinus alpinus* (L.)) in Loch Rannoch, Scotland.

Biological Conservation 43, 43-61. **Wallace, C.R. (1967).** Observations on the reproductive behaviour of the black bullhead (*Ictalurus melas*). *Copeia* 1967, 852-853. **Water Resources Board & Scottish Development Department (1974).** *The Surface Water Year Book of Great Britain, 1966–1970.* London: Her Majesty's Stationery Office. **Weatherley, A.H. (1959).** Some features of the biology of the Tench *Tinca tinca* (Linnaeus) in Tasmania. *Journal of Animal Ecology* 28, 73-87. **Weatherley, A.H. (1962).** Notes on the distribution, taxonomy and behaviour of tench *Tinca tinca* (L.) in Tasmania. *Annals and Magazine of Natural History* 4, 713-719. **Weatherley, N.S. (1987).** The diet and growth of the 0-group dace, *Leuciscus leuciscus* (L.), and roach, *Rutilus rutilus* (L.), in a lowland river. *Journal of Fish Biology* 30, 237-248. **Went, A.E.J. (1953).** The status of the shads, *Alosa finta* and *Alosa alosa* Cuvier, in Irish waters. *Irish Naturalists' Journal* 11, 8-11. **Went, A.E.J. (1971).** The distribution of Irish char, *Salvelinus alpinus*. *Irish Fisheries Investigations* 6A, 5-11. **Went, A.E.J. (1979).** 'Ferox' Trout, *Salmo trutta* L. of Lough Mask and Corrib. *Journal of Fish Biology* 15, 255-262. **Western, J.R.H. (1971).** Feeding and digestion in two cottid fishes, the freshwater *Cottus gobio* L. and the marine *Enophrys bubalis* (Euphrasen). *Journal of Fish Biology* 3, 225-246. **Wheeler, A. (1969).** *The Fishes of the British Isles and North West Europe.* Macmillan, London. **Wheeler, A. (1976).** On the populations of Roach (*Rutilus rutilus*), Rudd (*Scardinius erythropthalmus*), and their hybrid in Esthwaite Water, with notes on the distinctions between them. *Journal of Fish Biology* 9, 391-400. **Wheeler, A. (1978a).** Hybrids of Bleak, *Alburnus alburnus*, and Chub, *Leuciscus cephalus* in English rivers. *Journal of Fish Biology* 13, 467-473. **Wheeler, A. (1978b).** *Ictalurus melas* (Rafinesque 1820) and *I. nebulosus* (Lesueur 1879): the North American catfishes in Europe. *Journal of Fish Biology* 12, 435-439. **Wheeler, A., Blacker, R.W. & Pirie, S.F. (1975).** Rare and little-known fishes in British seas in 1970 and 1971. *Journal of Fish Biology* 7, 183-202. **Wheeler, A. & Easton, K. (1978).** Hybrids of Chub and Roach (*Leuciscus cephalus* and *Rutilus rutilus*) in English rivers. *Journal of Fish Biology* 12, 167-171. **Wheeler, A. & Jordan, D.R. (1990).** The status of the barbel, *Barbus barbus* (L.) (Teleostei, Cyprinidae), in the United Kingdom. *Journal of Fish Biology* 37, 393-400. **Wheeler, A. & Maitland, P.S. (1973).** The scarcer freshwater fishes of the British Isles. I. Introduced species. *Journal of Fish Biology* 5, 1-68. **Whilde, A. (1993).** *Threatened Mammals, Birds, Amphibians and Fish in Ireland: Irish Red Data Book 2: Vertebrates.* HMSO, Belfast. **Whoriskey, F.G., Fitzgerald, G.J. & Reebs, S.G. (1986).** The breeding season population structure of three sympatric, territorial sticklebacks (Pisces: Gasterosteidae). *Journal of Fish Biology* 29, 635-648. **Wildekamp, R.H., Van Neer, W., Küçük, F. & Ünlüsayin, M. (1997).** First record of the eastern Asiatic gobionid fish *Pseudorasbora parva* from the Asiatic part of Turkey. *Journal of Fish Biology* 51, 858-861. **Williams, W.P. (1965).** The population density of four species of freshwater fish, Roach, Bleak, Dace and Perch in the River Thames at Reading. *Journal of Animal Ecology* 34, 173-185. **Williams, W.P. (1967).** The growth and mortality of four species of fish in the River Thames at Reading. *Journal of Animal Ecology* 66, 695-720. **Williot, P. (Ed.) (1991).** *Acipenser.* CEMAGREF, Bordeaux. **Williot, P., Rochard, E., Castelnaud, G., Rouault, T., Brun, R., Lepage, M. & Elie, P. (1997).** Biological characteristics of European Atlantic Sturgeon, *Acipenser sturio*, as the basis for a restoration program in France. *Environmental Biology of Fishes* 48, 359-370. **Willoughby, L.G. (1970).** Mycological aspects of a disease of young perch in Windermere. *Journal of Fish Biology* 2, 113-116. **Wilson, J.P.F. (1983).** Gear selectivity, mortality and fluctuations in abundance of the Pollan *Coregonus autumnalis pollan* Thompson of Lough Neagh, Northern Ireland. *Proceedings of the Royal Irish Academy* 83B, 301-307. **Wilson, J.P.F. (1984).** The food of Pollan, *Coregonus autumnalis pollan* Thompson, in Lough Neagh, Northern Ireland. *Journal of Fish Biology* 24, 253-262. **Wilson, J.P.F. & Pitcher, T.J. (1983).** The seasonal cycle of condition in the Pollan, *Coregonus autumnalis pollan* Thompson, of Lough Neagh, Northern Ireland. *Journal of Fish Biology* 23, 365-370. **Wilson, J.P.F. & Pitcher, T.J. (1984).** Age determination and growth of the Pollan, *Coregonus autumnalis pollan* Thompson, of Lough Neagh, Northern Ireland. *Journal of Fish Biology* 24, 151-164. **Winfield, I.J. (1986).** The influence of stimulated aquatic macrophytes on the zooplankton consumption rate of juvenile roach *Rutilus rutilus*, rudd *Scardinius erythrophthalmus*, and perch *Perca fluviatilis*. *Journal of Fish Biology* 29A, 37-48. **Winfield, I.J., Bean, C.W. & Hewitt, D.P. (2002a).** The relationship between spatial distribution and diet of Arctic charr (*Salvelinus alpinus*) in Loch Ness, U.K. *Environmental Biology of Fishes* 64, 63-73. **Winfield, I.J., Cragg-Hine, D., Fletcher, J.M. & Cubby, P.R. (1996).** The conservation ecology of *Coregonus albula* and *C. lavaretus* in England and Wales, U.K. Pages 213-223 in: Kirchhofer, A. & Hefti, D. (Eds) *Conservation of Endangered Fish in Europe*. Birhauser Verlag, Basel. **Winfield, I.J., Crawshaw, D.H. & Durie, N.C. (2003).** Management of the cormorant, *Phalacrocorax carbo*, and endangered whitefish, *Coregonus lavaretus*, populations of Haweswater, UK. Pages 335-344 in: Cowx, I.G. (Ed.) *Interactions between Fish and Birds: Implications for Management*. Fishing News Books, Oxford. **Winfield, I.J., Fletcher, J.M. & Cubby, P.R. (1993a).** Confirmation of the presence of schelly, *Coregonus lavaretus*, in Brotherswater, UK. *Journal of Fish Biology* 42, 621-622. **Winfield, I.J., Fletcher, J.M. & Cubby, P.R. (1998).** The impact on the whitefish (*Coregonus lavaretus* (L.)) of reservoir operations at Haweswater, U.K. *Advances in Limnology* 50, 185-195. **Winfield, I.J., Fletcher, J.M. & James, J.B. (2004).** Conservation ecology of the vendace (*Coregonus albula*) in Bassenthwaite Lake and Derwent Water, U.K. *Annales Zoologici Fennici* **[In the press]**. **Winfield, I.J., Fletcher, J.M. & Winfield, D.K. (2002b).** Conservation of the endangered whitefish (*Coregonus lavaretus*) population of Haweswater, UK. Pages 232-241 in: Cowx, I.G. (Ed.) *Management and Ecology of Lake and Reservoir Fisheries*. Fishing News Books, Oxford. **Winfield, I.J., Tobin, C.M. & Montgomery, C.R. (1993b).** Ecological studies of the fish community. Pages 451-471 in: Wood, R.B. & Smith, R.V. (Eds) *Lough Neagh*. Kluwer, Dordrecht. **Winfield, I.J., Winfield, D.K. & Tobin, C.M. (1992).** Interactions between the roach, *Rutilus rutilus*, and waterfowl populations of Lough Neagh, Northern Ireland. *Environmental Biology of Fishes* 33, 207-214. **Winfield, I.J. & Wood, R.B. (1990).** Conservation of the Irish pollan, *Coregonus autumnalis pollan* Thompson, in Lough Neagh, Northern Ireland. *Journal of Fish Biology* 37A, 259-260. **Withler, I.L. (1966).** Variability in life history characteristics of steelhead trout (*Salmo gairdneri*) along the Pacific coast of North America. *Journal of the Fisheries Research Board of Canada* 23, 365-393. **Wood, A.B. & Jordan, D.R. (1987).** Fertility of roach x bream hybrids, *Rutilus rutilus* (L.) x

Abramis brama (L.), and their identification. *Journal of Fish Biology* 30, 249-262. **Woolland, J.V. (1987).** Grayling in the Welsh Dee: age and growth. *Journal of the Grayling Society* 1987, 33-38. **Woolland, J.V. & Jones, J.W. (1975).** Studies on Grayling, *Thymallus thymallus* L., in Llyn Tegid and the upper River Dee, North Wales. *Journal of Fish Biology* 7, 749-773. **Wootton, R.J. (1976).** *The Biology of the Sticklebacks*. Academic Press, London. **Wootton, R.J. (1984).** *The Functional Biology of Sticklebacks*. Croom Helm, Beckenham. **Wootton, R.J. & Mills, L.A. (1979).** Annual cycle in female Minnows *Phoxinus phoxinus* (L.) from an upland Welsh lake. *Journal of Fish Biology* 14, 607-618. **Worthington, E.B. (1941).** Rainbow Trout in Britain. *.Salmon and Trout Magazine* 100, 241-260; 101, 62-99. **Wright, P.J. & Huntingford, F.A. (1993).** Daily growth increments in the otoliths of the three-spined stickleback, *Gasterosteus aculeatus* L. *Journal of Fish Biology* 42, 65-78. **Wright, R.M. (1990a).** Aspects of the ecology of bream, *Abramis brama* (L.), in a gravel pit lake and the effects of reducing the population density. *Journal of Fish Biology* 37, 629-634. **Wright, R.M. (1990b).** The population biology of pike, *Esox lucius* L., in two gravel pit lakes, with special reference to early life history. *Journal of Fish Biology* 36, 215-230. **Wurtsbaugh, W.A., Brocksen, R.W. & Goldman, C.R. (1975).** Food and distribution of underyearling Brook and Rainbow Trout in Castle Lake California. *Transactions of the American Fisheries Society* 104, 88-95. **Wyatt, R.J. (1988).** The cause of extreme year class variation in a population of roach, *Rutilus rutilus* L., from a eutrophic lake in southern England. *Journal of Fish Biology* 32, 409-422.

Xie, S., Cui, Y. & Li, Z (2001). Dietary-morphological relationships of fishes in Liangzi Lake, China. *Journal of Fish Biology* 58, 1714-1729.

Yarrell, W. (1859). *A History of British Fishes*. Van Voorst, London. **Youngson, A.F., Knox, D. & Johnstone, R. (1992).** Wild adult hybrids of *Salmo salar* L. and *Salmo trutta* L. *Journal of Fish Biology* 40, 817-820.

Ziukov, M. & Petrova, G. (1993). On the pattern of correlation between the fecundity, length, weight and age of the pikeperch *Stizostedion lucioperca*. *Journal of Fish Biology* 43, 173-182.

INDEX TO COMMON AND SCIENTIFIC NAMES

Page numbers in **bold** refer to illustrations, and numbers in *italics* refer to distribution maps. Scientific names in brackets *[ ]* indicate well-known former names that have changed recently.

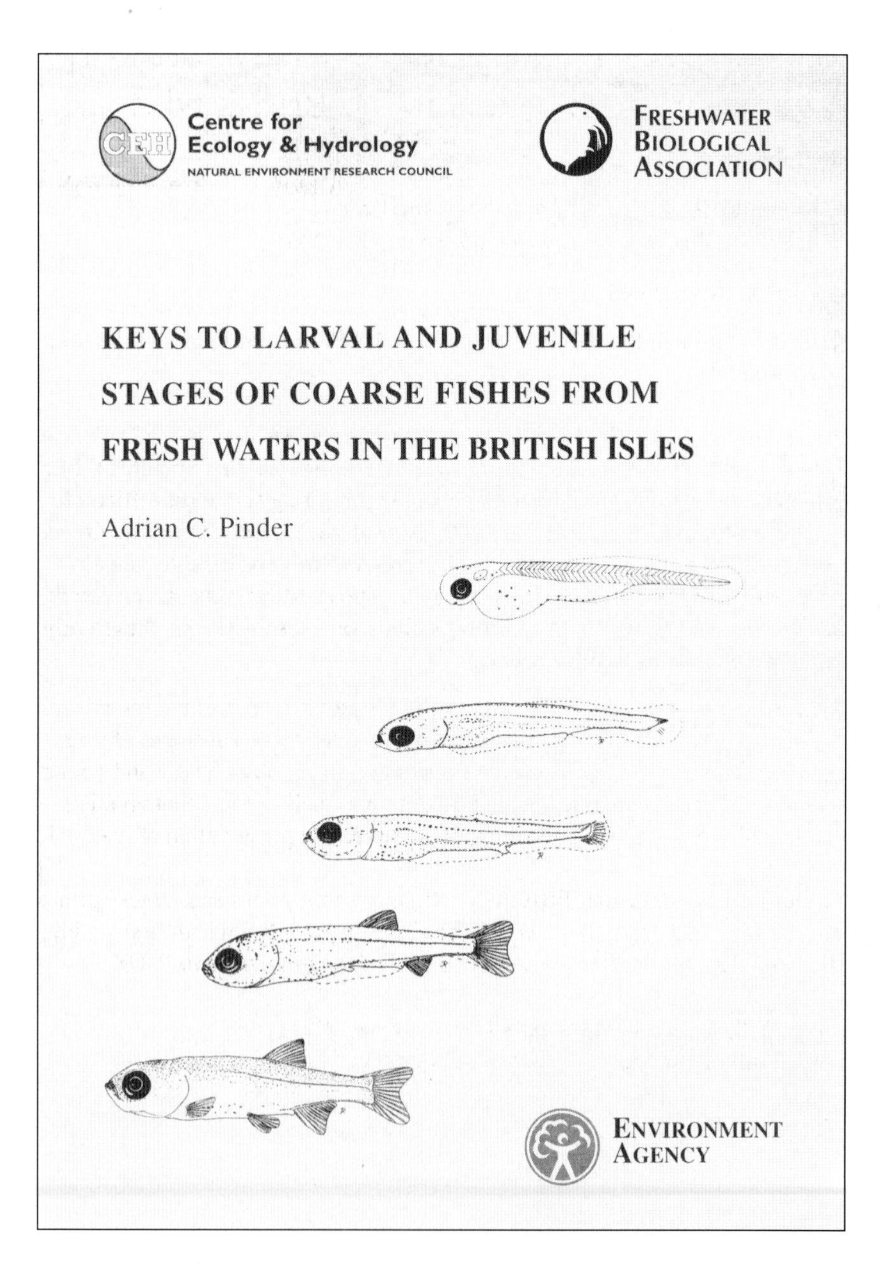
Centre for
Ecology & Hydrology
NATURAL ENVIRONMENT RESEARCH COUNCIL
FRESHWATER
BIOLOGICAL
ASSOCIATION
KEYS TO LARVAL AND JUVENILE
STAGES OF COARSE FISHES FROM
FRESH WATERS IN THE BRITISH ISLES
Adrian C. Pinder
ENVIRONMENT
AGENCY

Keys to larval and juvenile stages of coarse fishes from fresh waters in the British Isles

by Adrian C. Pinder

FRESHWATER BIOLOGICAL ASSOCIATION
SCIENTIFIC PUBLICATION NO. 60, September 2001

136 pages. ISBN 0 900386 67 3

Price £20.00, including packing & postage (FBA Members are entitled to a 25% discount)

This publication provides a series of keys to the early stages of "Coarse Fishes" that occur in British fresh waters. During the first month of free existence, these fishes develop through a number of stages that are difficult for aquatic biologists (and most ichthyologists) to recognise. This difficulty in distinguishing one species from another has led to considerable neglect of these early stages and their impact on the surrounding biota as predators, competitors and prey. Until recently, studies on recruitment in these early stages were also few and far between.

The series of keys produced by Adrian Pinder provides a most necessary tool to assist such studies. It comprises keys to five developmental stages of coarse fish: free embryos, young larvae, intermediate larvae, older larvae and young juveniles. Clearly illustrated with line-drawings and colour photographs, it also includes notes on the collection, preparation and preservation of material.

Further information about FBA Publications, current prices and Order Forms, may be obtained from Dept DWS, The Freshwater Biological Association, The Ferry House, Far Sawrey, AMBLESIDE, Cumbria LA22 0LP, UK.

Tel: +44 (0) 15394 42468. Fax: +44 (0) 15394 46914.
E-mail: info@fba.org.uk. Web: www.fba.org.uk.

WINDERMERE:
RESTORING THE HEALTH OF ENGLAND'S LARGEST LAKE

by Alan D. Pickering

FRESHWATER BIOLOGICAL ASSOCIATION
SPECIAL PUBLICATION NO. 11

Published by The Freshwater Biological Association,
Ambleside, December 2001

on behalf of The Lake District Still Waters Partnership:
Centre for Ecology and Hydrology, Freshwater Biological Association, Lake District National Park Authority, English Nature, Environment Agency, National Trust, United Utilities

126 + x pages, with 75 colour & black-and-white illustrations
ISBN 0 900386 68 1

Price £10.00, including packing & postage
(FBA Members & Booksellers are entitled to a 25% discount)

This book has been produced to celebrate the 50th Anniversary of the Lake District National Park (LDNP) and to inform readers about major issues of public interest concerning Windermere – the largest natural lake in England. The LDNP Authority and its Still Waters Partners are jointly responsible for overseeing and managing this national resource. They have to balance a wide variety of view-points and sometimes conflicting requirements against the natural 'needs' of Windermere itself, with the aim of maintaining the lake in an ecologically healthy state for public use and enjoyment now and in the future.

In this account, Alan Pickering starts by providing a background to Windermere, introducing the reader to the lake's geography, geology and historical development. As with all of the lakes in the English Lake District, and most of the UK, Windermere was formed at the end of the last glaciation. Major changes in its catchment induced by the changing climate and then by man's removal of the forests have been recognised through examination of lake sediments. The book concentrates on one of the most dramatic changes in the second half of the twentieth century – the decline in the health of the lake ecosystem, including the populations of Arctic Charr. The main cause of this was the increased nutrient load from local sewage treatment works. As the ecological balance of the lake was threatened, the collaborative efforts of scientists and environmental managers restored the lake to one with thriving populations of fish and other aquatic fauna and flora. As one of the scientists involved in this work, Professor Pickering gives a very clear account of all of these changes and some thoughts on issues surrounding the lake's future management. His clear explanations of the more technical aspects make it accessible for the less technically-minded reader as well as to scientists and students of lakes and their catchments.